Chambers
Adult Learners'
Guide to Numeracy

Geoff Mainwaring

Chambers

CHAMBERS

An imprint of Chambers Harrap Publishers Ltd

7 Hopetoun Crescent

Edinburgh, EH7 4AY

First published by Chambers Harrap Publishers Ltd 2007

Reprinted 2008

© Chambers Harrap Publishers Ltd 2007

A CIP catalogue record for this book is available from the British Library.

ISBN 978 0550 10321 5

Editor: Katie Brooks

Managing Editor: Mary O'Neill

Prepress: Heather Macpherson

Illustrations: Richard Duszczak

Designed and typeset by Chambers Harrap Publishers Ltd, Edinburgh

Printed by Legoprint, Italy

Contents

About the author

Geoff Mainwaring is a Learning Support and Numeracy Lecturer at Jewel and Esk Valley College. He has over 30 years' experience as a lecturer and tutor of mathematics and numeracy.

Acknowledgements

The author would like to thank Katie Brooks, editor at Chambers Harrap, for her invaluable support and guidance during the writing of this book; Karen Buist for being there to advise when words failed him; and the management and staff at Jewel and Esk Valley College.

Introduction

About this book

Chambers Adult Learners' Guide to Numeracy has been written especially for adults who want to improve their basic numeracy skills.

Many adults lack confidence about working with numbers, and find that their difficulties with numeracy can be an obstacle in their work, hobbies or everyday activities. If that description fits you, then this book will help you to build up your numeracy skills and show you how to apply them in practical situations.

The appearance and layout of the book have been designed to guide you through the necessary steps to improve your knowledge and understanding of the topics. Colour is used to show important steps in the calculations, and to highlight key words. You will also find the key words in the index, so that you can look them up easily. The book is written in simple, straightforward language and there are plenty of practice exercises to try out what you have learned.

The content of the book has been carefully selected to tie in with the 'Skills for Life' programme run by the Department for Education and Skills, which forms the basis for many adult numeracy courses. Each topic covers the basics before moving on to more advanced techniques.

These features mean that the book is suitable for you to use either as a self-study tool or in combination with an adult numeracy course.

Before you start

Before you start working with this book, it would be useful to spend some time thinking about your reasons for wanting to improve your numeracy skills. People tend to learn more easily if they have clear goals. Try making a list of areas of your life that will benefit by your improving your numeracy skills. For example, you might want to study numeracy because you would like to help your children with their homework. You might have a new job in mind that requires certain numeracy skills. Or you might simply want to revisit a subject that you feel you could have done better in at school. Identifying your goals before you start will give you a sense of purpose and help you focus.

Next, think about numeracy skills you already have – and you *will* have some! Everyone uses numbers in their everyday life. If you are a darts player then you may already be good at adding and subtracting. If you shop regularly then you are probably quite skilled at using money. If you are keen on DIY then you will be able to measure lengths accurately. Identifying what you are already good at will not only boost your confidence, but will also provide you with a foundation on which you can build.

Remember that you are in charge of your learning. Make the material in this book work for *you*. There is a section on Learning Styles towards the end of the book. Use it to try to identify the strategies that will help you learn. You can adapt the material in the book to suit your particular style(s). For example, to remember a rule you could write down an explanation, draw a diagram, or even make up a song!

How to use this book

Chambers Adult Learners' Guide to Numeracy is divided up into twelve sections, each covering a different topic. These are listed in the contents pages. Each section contains explanations of key skills for that topic area, worked examples, and practice exercises. At the back of the book you will find answers to the practice exercises, a section on Learning Styles and some multiplication tables. There are also some handy tables for converting between metric and imperial measures.

The book features some special symbols to highlight useful information:

 The 'light bulb' symbol indicates a handy tip or method

 The 'information' symbol indicates some useful information

 The 'warning' symbol indicates a warning about common mistakes

 The 'calculator' symbol indicates advice about using a calculator

You might not need to work through each section of the book. If, for example, you are confident using whole numbers then you might decide to start with the fractions section. But remember that numeracy is a subject where knowledge is built up step by step. It is difficult to understand percentages without a good working knowledge of fractions. It is difficult to understand the metric system without knowing how the decimal system works. A thorough understanding of each 'step' is important.

Learning number skills is not just a matter of remembering methods or rules. Although those things are important, a thorough understanding of the concepts behind the methods and rules is even more important. When working through the different sections of the book, don't be in too much of a hurry to move from one section to the next. Make sure you fully understand the work in one section before moving on. People often find difficulties with fractions, for example. This is usually because they don't understand where they come from and how they work. Time taken thinking about the concepts behind the methods and the rules will be well spent. Learning the methods and the rules will be much easier as a result and you will be less likely to forget them.

The practice exercises in each section of the book will give you lots of opportunity to test your knowledge and understanding. However, the more practice you do the better. If you are working with a tutor then he or she will probably be able to provide you with extra exercises. If not, then there are numerous websites where you will find downloadable and interactive materials. In the appendices to this book you will find a list of some current useful websites. Enter the name of any numeracy topic into an Internet search engine and you will find many more.

Some final advice

One of the common complaints from people studying numeracy is that they quickly forget what they have just learned. This is a normal reaction to learning things that will not be used regularly. Try to practise your new numeracy skills regularly – even just a half-hour a day. This will help you to retain the skills. Good luck!

WHOLE NUMBERS

What are whole numbers?

Whole numbers are the numbers we use to count in everyday life.

 Whole numbers are sometimes called **integers, counting numbers**, or **natural numbers**.

A brief history of numbers

Three million years ago, on a cave wall in Africa, a primitive hand recorded the number of people belonging to his or her tribe. The numbers are simple scratches on the rock, each scratch representing one person.

If you count the scratches you find that this tribe had 26 members. If there had been many more, then this early mathematician might have needed a bigger cave wall to record them all. A number system in which one mark represents one object has its limits. To record very large numbers you would need a very large piece of paper or a very large cave wall.

Early man partly got around this problem by collecting the marks in groups of five or ten, using a simple tally system that is still used today.

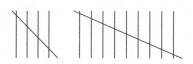

Whole numbers

Five thousand years ago, the Ancient Egyptians improved on this idea by writing symbols on papyrus to represent the numbers ten, hundred, thousand, and so on.

I	∩	☻	⚑
1	10	100	1000

Instead of 10 individual marks they wrote ∩

Instead of 100 individual marks they wrote ☻

So, the number 361 would be written as ☻ ☻ ☻ ∩∩∩∩∩∩ I

Similar developments were happening around this time in other parts of the world. The early Babylonian civilization also used symbols for groups of numbers. The Ancient Egyptians counted in tens but the Babylonians counted in sixes (which is why we have 60 seconds in a minute and 60 minutes in an hour).

This introduction of symbols represented a giant step forward in the use of numbers, but it still had its limits. For very large numbers you would still need a large sheet of paper or papyrus.

The later Babylonians, along with the Chinese, the Mayans and the Hindus, got round this problem by introducing the concept of **place value**. Look back at the Ancient Egyptian number representing 361. The order the symbols appeared in did not matter to the Egyptians. They could be written in a circle, or one on top of another. What the Babylonians proposed was revolutionary: the *position* of a symbol in a number could change its value.

Two thousand years later, in India, mathematicians extended the idea of having symbols for some of the numbers by using a different symbol for *each* figure in their numbering system. This development, combined with the idea of place value, led to the gradual evolution of most modern number systems.

۹	২	३	৪	५	६	७	⊏	৫	০
1	2	3	4	5	6	7	8	9	0

Notice the similarity between these early Indian numbers and the modern numbers we use today.

Place value

The numbers that we use are made up from just ten symbols, called figures or digits.

0 1 2 3 4 5 6 7 8 9

We are able to write so many different numbers with only ten symbols because the **position** that the symbol occupies in the number affects its **value**.

For example, the number 123 (one hundred and twenty three) is different from the number 321 (three hundred and twenty one) even though the same figures are being used. In the second number the figures are written in a different order. The different positions of the figures create a different number.

This is how it works:

The first figure on the right hand side of the number represents the number of ones (1s), or units. The figure next to it, moving to the left, represents the number of tens (10s), the figure next to *that* shows the number of hundreds (100s).

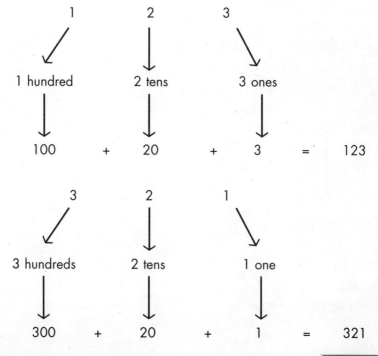

Whole numbers

As we move to the left, the place value of each figure is ten times bigger than the place value of the figure before it. We can extend this idea of place value without limit.

10000s	1000s	100s	10s	1s
(ten thousands)	(thousands)	(hundreds)	(tens)	(ones)

← The columns don't stop here. They can be extended indefinitely.

Any number can be broken down in this way, but in practice we do not have to do this because we learn to recognize most numbers.

Large numbers

The reason why very large numbers are sometimes confusing is that they are unfamiliar to us – we haven't learned to recognize them.

Given a number like 74914, it might be difficult to say what it is. However, by using the idea of place value we can work it out.

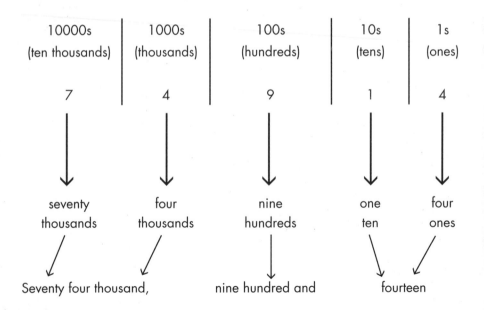

Usually, we do not have to break numbers down in this way, so don't worry if you find this a bit confusing. However, try to remember the underlying concepts that our number system is based on groups of ten, and that the value of the figures in all our whole numbers, reading from right to left, correspond to 1s, 10s, 100s, 1000s, and so on.

Sometimes people write large numbers with spaces or commas between each group of three figures so that they are easier to read. In later sections of this book, thin spaces are used to break up large numbers.

74,914   74914

Using numbers – useful skills

In everyday life it is useful to be able to add amounts together, whether they are amounts of money, or lengths when doing a spot of DIY, or the total number of miles to be driven on holiday, etc. **Addition (+)** is a useful skill.

It is also useful to take one amount from another. For example, you might want to deduct tax from your salary to find your net pay. **Subtraction (–)** is a useful skill.

In some situations you may find that you have to add the same amount a number of times. For example, you might know the cost of one cinema ticket, and want to work out how much it will cost for a group of people to go. **Multiplication (×)** allows you to do this quickly.

Similarly, you might find you have to take away the same amount a number of times, or split something into equal shares. For example, if you go out for a meal with friends, you may want to split the bill into equal shares to work out how much each person should pay. **Division (÷)** allows you to do this quickly.

These four skills **+ – × ÷** are called **operations**.

Addition

Other words that mean the same as addition are: **add, adding, finding the total, finding the sum, plus, more than**.

We find the total or sum of two or more numbers by adding them together. This can be done in a number of ways:

Counting on

> ### Example
>
> An adult numeracy evening class has 6 students attending. 3 more students join the class. How many students are there now?

We can find the new total by 'counting on' from 6 until we have added 3 more.

$$6 \rightarrow 7 \rightarrow 8 \rightarrow 9 \qquad \text{so } 6 + 3 = 9$$

3 more

You can count on 'in your head', by saying the numbers out loud, or by using your fingers. Using fingers is fine – fingers were the world's first calculators.

Using a number line

A number line can also be used to do this.

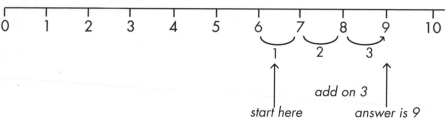

start here

add on 3

answer is 9

A ruler can be used as a simple number line.

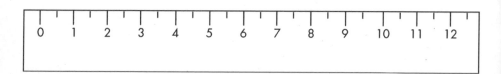

Number bonds

With practice and use we get to remember the answers when we add certain pairs of numbers together.

For example:

$$2 + 3 = 5$$
$$5 + 5 = 10$$
$$4 + 6 = 10$$

These combinations are sometimes referred to as **number bonds**. The more of these we can remember, the quicker addition becomes. The best way to learn and remember these combinations is by practice and use rather than by trying to memorize them.

 Number bonds are reversible: $2 + 3 = 5$ and $3 + 2 = 5$.

Formal addition

With bigger numbers, counting on becomes a long process. It is useful to be able to set out addition as in the following example:

Example

The distance from Edinburgh to Newcastle is 170 miles. The distance from Newcastle to Birmingham is 126 miles. How far is it from Edinburgh to Birmingham?

To find the total distance, we have to add the two distances together.

170 miles + 126 miles

Adding these by 'counting on' would be a slow process. Instead, we put the two numbers under each other, making sure the ones are above the ones, the tens above the tens and the hundreds above the hundreds.

$$
\begin{array}{r}
1\,7\,0 \\
+\ 1\,2\,6 \\
\end{array}
$$

First, add the ones, then the tens, and, finally, the hundreds

$$
\begin{array}{r}
1\,7\,0 \\
+\ 1\,2\,6 \\
\hline
6 \\
\end{array}
\qquad
\begin{array}{r}
1\,7\,0 \\
+\ 1\,2\,6 \\
\hline
9\,6 \\
\end{array}
\qquad
\begin{array}{r}
1\,7\,0 \\
+\ 1\,2\,6 \\
\hline
2\,9\,6 \\
\end{array}
$$

Here the steps of the addition have, again, been shown as separate calculations. In practice, the three steps can be done in the one calculation.

$$
\begin{array}{r}
1\,7\,0 \\
+\ \underline{1\,2\,6} \\
2\,9\,6
\end{array}
$$

The final answer is 296 miles.

Practice 1

Try adding these pairs of numbers using this method. The answers are in the back of the book when you are ready to check.

1.	32 + 12	6.	17 + 121
2.	56 + 41	7.	954 + 23
3.	120 + 54	8.	1205 + 83
4.	236 + 123	9.	4750 + 149
5.	400 + 456	10.	128 457 + 231 532

You will notice that in the last three questions the numbers have become quite large. The advantage of adding like this is that it doesn't matter how large the numbers are – even millions! – you can still add them step by step.

 Be careful with zeros (0). If you add a zero (0) to another number, you are adding nothing so the number remains unchanged.

6 + 0 = 6

3 + 0 = 3

Addition with carrying

Sometimes, adding two numbers produces an answer that is bigger than the place value of the numbers. If this sounds a bit complicated, the following example might make it clearer.

Example

Jane had saved £16 towards a bicycle. On her birthday she received another £18 pounds. How much did she now have in total?

In this example we need to add together £16 and £18

$$\begin{array}{r} 1\ 6 \\ +\ \underline{1\ 8} \end{array}$$

Adding the first column:

$$6 + 8 = 14$$

But the numbers in the first column are ones – they only go up to 9. We can't squeeze 14 into the space in the answer below the 6 and the 8 – there is only room for one figure. Instead, we remember that 14 is made up of one 10 and four 1s. We put the 4 of the 14 (the four ones) below the 6 and the 8 and 'carry' the 1 of the 14 (the ten) into the next column. So, we are taking 10 of the 14 and moving it into the tens column.

$$\begin{array}{r} 1\ 6 \\ +\ \underline{1\ 8} \\ 4 \\ 1 \leftarrow \end{array}$$

When we add the numbers in the next column we have to remember the 1 that we 'carried'. That is now in the tens column and has to be added to whatever other figures are in that column.

$$\begin{array}{r} 1\ 6 \\ +\ \underline{1\ 8} \\ 3\ 4 \\ 1 \leftarrow \end{array}$$

So Jane has a total of £34.

You can show the figure that you carried at the bottom of the calculation, as shown above, or at the top. Nowadays, schools tend to encourage students to show the number at the top.

$$
\begin{array}{r}
1 \\
1\,6 \\
+\ \underline{1\,8} \\
3\,4
\end{array}
$$

This idea of carrying to the next column should be used whenever the answer for one column becomes more than 9. In the next example this happens with the tens column:

Example

Calculate 68 + 41.

$$
\begin{array}{r}
6\,8 \\
+\ \underline{4\,1}
\end{array}
$$

First column: 8 + 1 = 9. This is not more than 9, so we can just put it in the answer.

$$
\begin{array}{r}
6\,8 \\
+\ \underline{4\,1} \\
9
\end{array}
$$

Second column: 6 + 4 = 10. This is more than 9, so we have to put a zero in the answer space below the 6 and the 4 and carry the 1 to the next column.

$$
\begin{array}{r}
6\,8 \\
+\ \underline{4\,1} \\
\underline{0\,9} \\
1 \leftarrow
\end{array}
$$

In this example, there are no figures in the 100s column, so to complete the addition we have to add the 1 we carried into the empty 100s column.

```
    6 8
  + 4 1
  1 0 9
```

The answer is 109.

In this example we are actually carrying 100, since the 6 and the 4 were tens (60 + 40 = 100). In practice, there is no need to worry about this. Just remember that, if an answer for any column gets bigger than 9, you carry the extra to the next column.

Don't be intimidated by large numbers. However large they are, they can still be added step by step.

Example

Calculate 2345 + 5798.

```
    2 3 4 5
  + 5 7 9 8
    8 1 4 3
    1 1 1
```

Practice 2

1.	39 + 12	6.	89 + 121
2.	56 + 35	7.	984 + 28
3.	128 + 56	8.	1809 + 373
4.	337 + 123	9.	4750 + 2576
5.	567 + 436	10.	25768 + 98676

Carrying larger figures

So far, in our examples, the 'extra' that has to be carried has always been a 1. When adding two numbers, the extra will always be a 1. If, however, you are adding more than two numbers then the extra to be carried could be a 2 or a 3, or anything up to 9, as shown in the next couple of examples:

Example

Calculate 37 + 56 + 78.

$$
\begin{array}{r}
3\,7 \\
5\,6 \\
+\ \underline{7\,8}
\end{array}
$$

Adding the first column:

$$7 + 6 + 8 = 21$$

We put down the 1 and carry the 2:

$$
\begin{array}{r}
3\,7 \\
5\,6 \\
+\ \underline{7\,8} \\
1 \\
2 \leftarrow
\end{array}
$$

Adding the second column: 3 + 5 + 7 + the 2 that has been carried:

$$3 + 5 + 7 + 2 = 17$$

$$
\begin{array}{r}
3\,7 \\
5\,6 \\
+\ \underline{7\,8} \\
7\,1 \\
1\,2
\end{array}
$$

The 1 hundred that we carried can now be added into the empty 100s column :

$$
\begin{array}{r}
37 \\
56 \\
+\ 78 \\
\hline
171 \\
12
\end{array}
$$

The final answer is 171.

Subtraction

 Other words that mean the same as subtraction are: **taking away**, **subtracting**, **finding the difference**, **minus**, **less than**.

We find the difference between two numbers by taking them away from each other. This is called subtraction. Like addition, it can be done in a number of ways.

Counting on

We can subtract one number from another by starting with the smaller number and counting on until we reach the larger.

Example

Mary's daughter is 5 years old. Her son is 9 years old. What is the difference in their ages?

Start with the smaller number, 5, and count on until you reach the larger, 9.

$$5 \rightarrow 6 \rightarrow 7 \rightarrow 8 \rightarrow 9$$

4 more

So 9 – 5 = 4.

As with addition, counting on can be done in your head or on your fingers. You can also use a number line or a ruler.

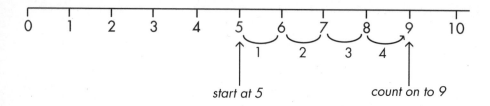

start at 5   *count on to 9*

Formal subtraction

As with addition, counting on becomes a long process with bigger numbers. It is useful to be able to set out subtraction as in the following example:

Example

Jasmin's car insurance costs her £253 a year. She wants to buy a new car but her insurance will rise to £366 a year. What would be the increase in her insurance if she bought the new car?

In this example we need to find the difference between £366 and £253. To do this we have to take £253 away from £366. We have to subtract one number from the other.

$$£366 - £253$$

Put the two numbers under each other as in addition, making sure the ones are above the ones, the tens above the tens and the hundreds above the hundreds.

 In addition it did not matter which number you put on the top and which on the bottom, but in subtraction it is very important that the number on the bottom is the number you want to take away.

```
  £ 3 6 6        ✓        £ 2 5 3        ✗
- £ 2 5 3               - £ 3 6 6
```

First, subtract the ones. Then, subtract the tens. Finally, subtract the hundreds.

```
  £ 3 6 6            £ 3 6 6            £ 3 6 6
- £ 2 5 3          - £ 2 5 3          - £ 2 5 3
        3                1 3              1 1 3
```

Here the steps of the subtraction have, again, been shown as separate calculations. In practice, the three steps can be done in the one calculation.

$$£366$$
$$-\ \underline{£253}$$
$$£113$$

The final answer is £113. Jasmin's car insurance will cost her £113 more each year if she buys a new car.

Practice 3

Try subtracting these pairs of numbers.

1. 36 – 21
2. 78 – 52
3. 89 – 35
4. 245 – 123
5. 467 – 22

6. 678 – 123
7. 4567 – 303
8. 7541 – 5331
9. 3464 – 300
10. 87 375 – 65 124

Be careful with zeros (0). If you subtract a zero (0) from another number, you are taking away nothing so the number remains unchanged.

5 – 0 = 5
9 – 0 = 9

Subtraction with borrowing

The following example shows a difficulty that sometimes arises with subtraction.

Example

I earn £382 a week taxable income and pay £56 a week income tax. What would be my weekly pay after tax has been deducted?

Whole numbers

To find what is left after tax has been deducted, we have to take £56 from £382. Setting this out as before:

$$£\ 3\ 8\ 2$$
$$-\ £\ \ \ 5\ 6$$

The difficulty arises when we try to subtract the 6 from the 2, because 6 is bigger than 2.

We can overcome this by 'borrowing' from the 8 in the tens column. We borrow one 10 and make it into ten 1s, leaving us with only seven 10s.

$$10$$
$$\rightarrow$$
$$£\ 3\ \overset{7}{8}\ 2$$
$$-\ £\ \ \ 5\ 6$$

We add the ten 1s to the 2 already in the ones column, making twelve 1s in total. The borrowing process can be written like this:

$$£\ 3\ \overset{7}{8}\ {}^{1}2$$
$$-\ £\ \ \ 5\ 6$$

We have 'cashed in' one of the tens and moved it into the ones column.

Now we can do the subtraction:

$$£\ 3\ \overset{7}{8}\ {}^{1}2$$
$$-\ £\ \ \ 5\ 6$$
$$£\ 3\ 2\ 6$$

My weekly pay would be £326.

This might seem a very complicated and lengthy procedure. In practice you don't have to worry about whether you are moving a ten out of the tens column or a hundred out of the hundreds column. Think in terms of moving 1 from one column to the next. Remember to reduce the figure you've borrowed from by 1.

Example

Calculate 239 – 167.

$$\overset{1}{\cancel{2}}\overset{1}{3}9$$
$$-\ \underline{1\ 6\ 7}$$
$$7\ 2$$

The answer is 72.

Practice 4

1.	57 – 38	6.	234 – 152
2.	94 – 39	7.	254 – 66
3.	22 – 18	8.	105 – 73
4.	145 – 57	9.	543 – 176
5.	283 – 38	10.	1456 – 278

Subtraction with repeated borrowing

Sometimes when we want to borrow in a subtraction calculation there seems to be nothing there to borrow from.

Example

What is 305 – 26?

$$3\ 0\ 5$$
$$-\ \ \underline{2\ 6}$$

Subtracting the units: we can't take 6 from 5, so we try to borrow from the tens column. However, there is a zero in the tens column. Before we can borrow from the tens column, we have to borrow from the hundreds column.

$$\overset{2}{\cancel{3}}\overset{1}{\cancel{0}}5$$
$$-\ \ \underline{2\ 6}$$

Whole numbers

Having borrowed from the hundreds column there is now enough in the tens column to borrow from:

$$9$$
$$^2\cancel{3}\,^1\cancel{0}\,^15$$
$$-\quad 2\ 6$$

The subtraction becomes:

$$2\ 9\,^15$$
$$-\quad 2\ 6$$
$$2\ 7\ 9$$

The answer is 279.

Multiplication

 Other words that mean the same as multiplication are: **multiplying, times**.

Multiplication is a quick way to calculate totals of the same number.

Example

Jane is buying pizzas for her son's birthday party. Each pizza costs £2 and she needs to buy 8 of them. What is the total cost?

This could be calculated by repeated addition:

£2 + £2 + £2 + £2 + £2 + £2 + £2 + £2 = £16

– here we are adding eight £2

Or by multiplication:

£2 × 8

The multiplication sign is used to represent repeated addition.

Many people, if asked what eight twos are, will add up the twos, but this defeats the object of multiplying as a quicker method.

To use multiplying instead of adding we need to know what 'eight twos' are.

Multiplying tables and multiplying squares

One way of knowing what eight twos, or three fives (5 + 5 + 5), or eight sevens (7 + 7 + 7 + 7 + 7 + 7 + 7 + 7) are, without adding them up, is to memorize the multiplication tables shown in the appendix. These give us the answers to all the combinations of numbers between 1 and 9.

You might have learned these tables at school. If you didn't, it's worth trying to do it now – the effort will be worth it in the long run. There are various ways of committing these tables to memory. The appendix contains a learning styles questionnaire that will help you to find the way that will suit you best.

However, if you find learning the tables very difficult, don't despair! The multiplying square shown below is a useful tool.

	1	2	3	4	5	6	7	8	9	10
1	1	2	3	4	5	6	7	8	9	10
2	2	4	6	8	10	12	14	16	18	20
3	3	6	9	12	15	18	21	24	27	30
4	4	8	12	16	20	24	28	32	36	40
5	5	10	15	20	25	30	35	40	45	50
6	6	12	18	24	30	36	42	48	54	60
7	7	14	21	28	35	42	49	56	63	70
8	8	16	24	32	40	48	56	64	72	80
9	9	18	27	36	45	54	63	72	81	90
10	10	20	30	40	50	60	70	80	90	100

Example

Karen buys 5 boxes of eggs. Each box contains 6 eggs. How many eggs does she have altogether?

We need to work out 6 × 5. Find 6 in the left hand column and 5 in the first row. Track across from the 6 until you are under the 5. The number you arrive at, 30, is the answer.

Practice 5

Using either the multiplying square or your knowledge of the multiplying tables, write down the answers to the following multiplications.

1.	5 × 6		6.	8 × 9
2.	7 × 6		7.	3 × 9
3.	3 × 8		8.	7 × 7
4.	5 × 9		9.	9 × 2
5.	4 × 7		10.	5 × 5

Multiplying larger numbers

We often need to multiply numbers that have more than one digit.

Example

It costs Joe £24 a month to rent a garage. How much does it cost him for a year (12 months)?

To answer the question we need to calculate what twelve twenty-fours are.

(£24 + £24 +£24 +£24 +£24 +£24 +£24 +£24 +£24 +£24 +£24 +£24)

Using multiplication rather than addition, this could be written:

£24 × 12

Because both these numbers are larger than 10, it is difficult to use tables or the multiplying square to get to the answer in one step – people don't usually know their twelve times or twenty-four times tables or have a multiplying square that big. Instead, we have to do multiplications like this in steps.

First write down the two numbers under each other like you did with addition and subtraction. Make sure that the ones are under each other and that the tens are under each other.

$$\begin{array}{r} 2\,4 \\ \times \quad \underline{1\,2} \end{array}$$

Now instead of trying to multiply by the 12 in one go, work through the following steps:

 First multiply by the 2 – get an answer.

 Next, multiply by the 1 – get an answer.

 Add the two answers together.

Multiplying by the 2:

$$\begin{array}{r} 2\,4 \\ \times\ \underline{1\,2} \\ 4\,8 \end{array}$$

Multiplying by the 1:

$$\begin{array}{r} 2\,4 \\ \times\ \underline{1\,2} \\ 4\,8 \\ 2\,4 \end{array}$$

Adding the two answers together:

$$\begin{array}{r} 2\,4 \\ \times\ \underline{1\,2} \\ 4\,8 \\ +\ \underline{2\,4\ } \\ 2\,8\,8 \end{array}$$

It costs Joe £288 to rent his garage for a year.

A couple of 'rules' to remember

Always start your multiplying with the 'ones'.

Always start writing your answer below the figure you are multiplying by – when multiplying by the 2, the 8 in that part of the answer is written under the 2; when multiplying by the 1, the 4 in that part of the answer is written under the 1. (This makes sure that the figures in the answer are in the right columns.)

If there are gaps, like the one under the 8, a zero can be put in, but this is not essential.

Tables or a multiplying square can be used to give the answers to the different parts of the multiplication.

One more example, this time using 'hundreds':

Example

Calculate 103 × 23.

```
      1 0 3
  ×     2 3
      3 0 9        – multiplying 103 by the 3
    2 0 6          – multiplying 103 by the 2
    2 3 6 9
```

Be careful with zeros (0). If you multiply by a zero (0) the answer is 0.

$0 \times 6 = 0$

When you multiply numbers, it doesn't matter which way round you do the multiplication; for example 6 x 5 is the same as 5 x 6. When you multiply large numbers, it is usually less confusing to multiply the larger number by the smaller one. For example, it is easier to calculate 259 x 3 than 3 x 259.

Practice 6

1.	45 × 12	6.	79 × 89
2.	23 × 16	7.	123 × 15
3.	18 × 13	8.	305 × 62
4.	50 × 21	9.	4351 × 65
5.	92 × 33	10.	12 × 1726

Multiplying by 10, 100, 1000

Ten (10) has one zero. To multiply a whole number by ten, add a zero to the whole number.

$$36 \times 10 = 360$$
$$1632 \times 10 = 16320$$

A hundred (100) has two zeros. To multiply a whole number by a hundred, add two zeros to the whole number.

$$347 \times 100 = 34700$$
$$21 \times 100 = 2100$$

A thousand (1000) has three zeros. To multiply a whole number by a thousand, add three zeros to the whole number.

$$245 \times 1000 = 245000$$
$$5 \times 1000 = 5000$$

Division

 Other words that mean the same as division are: **dividing, sharing**.

Division is the reverse of multiplication. Whereas multiplication gave us a quick method of repeated addition, division gives us a quick method of repeated subtraction. The following example illustrates this:

Example

Samantha has £20 in her piggy bank. Each week she takes out £5 to spend. How many weeks before her piggy bank is empty?

The calculation could be done like this:

1st week:	£20 – £5	=	£15
2nd week:	£15 – £5	=	£10
3rd week:	£10 – £5	=	£5
4th week:	£5 – £5	=	0

So Samantha was able to take out £5 pounds for 4 weeks before her piggy bank was empty.

Another way of thinking about this is to ask the question:

'How many lots of £5 are in £20?'

or:

'How many lots of £5 would you have to add together to make £20?'

or:

'What would you have to multiply £5 by to get £20?'

If we know the multiplying tables then we know that 4 times 5 is 20; four fives make twenty (see the section about multiplication on page 18). So, by using our knowledge of the multiplication tables, we can work out how many fives there are in twenty.

When we calculate how many times a number goes into another number we are dividing the one number by the other. This is written using the division sign ÷ as

$$20 \div 5 = 4$$

Using the multiplying square for dividing

If you feel you do not know your multiplication tables well enough, then you can use the multiplying square to do division.

Example

What is 40 ÷ 5?

This question means the same as 'how many times does 5 go into 40?' or 'what would 5 have to be multiplied by to give 40?'

Look at the multiplying square and follow the arrows. Can you see how the square helps us find the answer?

	1	2	3	4	5	6	7	8	9	10
1	1	2	3	4	5	6	7	8	9	10
2	2	4	6	8	10	12	14	16	18	20
3	3	6	9	12	15	18	21	24	27	30
4	4	8	12	16	20	24	28	32	36	40
5	5	10	15	20	25	30	35	40	45	50
6	6	12	18	24	30	36	42	48	54	60
7	7	14	21	28	35	42	49	56	63	70
8	8	16	24	32	40	48	56	64	72	80
9	9	18	27	36	45	54	63	72	81	90
10	10	20	30	40	50	60	70	80	90	100

Find 5 in the left hand column. Track across until you find 40. Then track up to the number in the top row to find the answer. The number in the top row is 8.

So, 40 ÷ 5 = 8.

Practice 7

Use the multiplying square, or your knowledge of the multiplying tables, to find the answers to the following divisions:

1. 20 ÷ 5
2. 64 ÷ 8
3. 24 ÷ 8
4. 18 ÷ 3
5. 70 ÷ 10

6. 56 ÷ 8
7. 42 ÷ 7
8. 50 ÷ 5
9. 49 ÷ 7
10. 54 ÷ 6

Dividing with remainders

Think about the following division:

$$10 \div 3$$

How many threes are in ten? Three threes are nine ($3 \times 3 = 9$) and there is 1 left over, or remaining. In some text books you may see divisions which produce remainders laid out like this:

$$\begin{array}{r} 3\,R1 \\ 3\overline{)1\ 0} \end{array}$$

In a later section in this book we will see how remainders like this can be expressed as fractions.

Dividing larger numbers

Example

> Gordon has a budget of £240 to buy text books for his evening class. If each book costs £8, how many can he buy?

In this example we want to know how many eights we can get out of 240. We need to answer the question

'How many times will 8 go into 240?'

or:

'What would 8 have to be multiplied by to give 240?'

Consulting the multiplication square on the next page (or the tables in our head) we see that our eight times table only goes up to 80 (ten eights are eighty).

Instead of trying to find out, in one go, how many times 8 will go into 240, we can break the division down into steps as shown below:

$$8\overline{)2\ 4\ 0}$$

Firstly, we ask how many times will 8 go into 2, the first figure of the 240. The answer is that 8 can't go into 2 because 2 is smaller than 8. (Later in this book we will see that, if we are allowed to consider fractional and decimal answers, then it is possible to divide 2 by 8. Here, however, we are dealing with whole numbers, so no answer is possible.)

	1	2	3	4	5	6	7	8	9	**10**
1	1	2	3	4	5	6	7	8	9	10
2	2	4	6	8	10	12	14	16	18	20
3	3	6	9	12	15	18	21	24	27	30
4	4	8	12	16	20	24	28	32	36	40
5	5	10	15	20	25	30	35	40	45	50
6	6	12	18	24	30	36	42	48	54	60
7	7	14	21	28	35	42	49	56	63	70
8	8	16	24	32	40	48	56	64	72	80
9	9	18	27	36	45	54	63	72	81	90
10	10	20	30	40	50	60	70	80	90	100

Next, we ask how many times will 8 go into 24, using the first and second figures of the 240. Another way of asking this question is to ask what we would have to multiply 8 by to get 24. Using the multiplying square, or our knowledge of the multiplication tables, we can see that the answer is 3.

	1	2	**3**	4	5	6	7	8	9	10
1	1	2	3	4	5	6	7	8	9	10
2	2	4	6	8	10	12	14	16	18	20
3	3	6	9	12	15	18	21	24	27	30
4	4	8	12	16	20	24	28	32	36	40
5	5	10	15	20	25	30	35	40	45	50
6	6	12	18	24	30	36	42	48	54	60
7	7	14	21	28	35	42	49	56	63	70
8	8	16	24	32	40	48	56	64	72	80
9	9	18	27	36	45	54	63	72	81	90
10	10	20	30	40	50	60	70	80	90	100

$$\overset{3}{8\,\overline{)2\ 4\ 0}}$$

The 3 is placed above the 4 as shown

Lastly, we ask how many times will 8 go into 0, the last figure of the 240. The answer to this must be 0, since you can't get any 8s out of 0.

$$\overset{3\ 0}{8\,\overline{)2\ 4\ 0}}$$

The 0 is placed above the 0 as shown

Notice that we have to put a zero in the answer to hold the 3 in its correct position. Compare this with when we divided the 2 by 8 at the beginning of the division. There was no need to put a zero in the answer there, because it didn't affect the position of the other figures in the answer.

There are no further figures to consider, so we have come to the end of the calculation. The answer to our original question – 'how many times does 8 go into 240?' – is 30.

Gordon can buy 30 books.

 Be careful with zeros. Dividing a zero by any other whole number gives an answer of zero.

Division with carrying

Sometimes we have to carry numbers during a division sum, just as we did with addition (see page 8).

> **Example**
>
> Calculate 952 ÷ 8.

We can set out this calculation as follows:

$$8\,\overline{)9\ 5\ 2}$$

Dividing 8 into 9:

8 goes into 9 once with 1 'left over'. The 1 'left over' in this example is actually 1 hundred. This can be changed into tens (1 hundred = 10 tens) and added to the 5 tens, making 15 tens in all. We say that the 1 left over has been 'carried' to the 5.

$$\begin{array}{r} 1 \\ 8\overline{\smash)9^15\ 2} \end{array}$$

Dividing 8 into 15:

8 goes into 15 once with 7 'left over' (15 – 8 = 7), so the 7 can be 'carried' to the next figure – the 2. This time, the 7 'left over' is 7 tens. This can be changed in 70 ones (or units) and added to the 2 ones, making 72 ones in all.

$$\begin{array}{r} 1\ 1 \\ 8\overline{\smash)9^15^72} \end{array}$$

Dividing 8 into 72:

8 goes into 72 nine times – use multiplying square or tables.

$$\begin{array}{r} 1\ 1\ 9 \\ 8\overline{\smash)9^15^72} \end{array}$$

So the final answer is:

$$952 \div 8 = 119$$

Practice 8

1.	1235 ÷ 5	6.	2100 ÷ 10
2.	1628 ÷ 4	7.	6634 ÷ 2
3.	208 ÷ 8	8.	1848 ÷ 8
4.	1730 ÷ 5	9.	243 ÷ 3
5.	558 ÷ 9	10.	8000 ÷ 5

Dividing by 10, 100, 1000

If a number ends in one or more zeros then we can sometimes divide by 10, 100 or 1000 by removing zeros from the end of the number as shown below.

Ten (10) has one zero. To divide a whole number by ten, remove a zero from the whole number.

$$30 \div 10 = 3$$
$$240 \div 10 = 24$$
$$500 \div 10 = 50$$

A hundred (100) has two zeros. To divide a whole number by a hundred, remove two zeros from the whole number.

$$400 \div 100 = 4$$
$$1000 \div 100 = 10$$
$$2\,438\,700 \div 100 = 24\,387$$

A thousand (1000) has three zeros. To divide a whole number by a thousand, remove three zeros from the whole number.

$$2000 \div 1000 = 2$$
$$15\,000 \div 1000 = 5$$
$$2\,000\,000 \div 1000 = 2000$$

Dividing by large numbers

In the following example, we are dividing by a number greater than 10, and this makes it difficult to use the multiplying square or tables.

Example

Calculate $3645 \div 15$.

We can set out the division as before:

$$15 \overline{)\,3645}$$

First, we work out how many times 15 goes into 3 – it won't go, so we carry the 3 to the next number.

Next, we work out how many times 15 goes into 36. The multiplying square does not show combinations of 15, so we have to do this in a different way.

Check if 15 will go into 36 once.

We know that 15 will go into 36 once because 36 is bigger than 15.

Find out what is left over.

$$36 - 15 = 21$$

Because the remainder is bigger than 15, this tells us that the 15 can go into 36 more than once.

Check if 15 will go twice.

$$15 \times 2 = 30$$

Find out what is left over.

$$36 - 30 = 6$$

This is less than 15, so we can't get another 15 out of it. Instead, we carry the six to the next figure.

$$15 \overline{\smash{\big)}\, 3\,6\,^6{4}\,5} \overset{2}{}$$

Repeating the process for 64:

We need to find out how many times 15 will go into 64.

$$1 \times 15 = 15$$
$$2 \times 15 = 30$$
$$3 \times 15 = 45$$
$$4 \times 15 = 60 \qquad \textit{15 will go 4 times}$$
$$5 \times 15 = 75$$

The remainder will be:

$$64 - 60 = 4$$

We carry the 4 to the next figure.

$$\begin{array}{r} 2\ 4 \\ 15\ \overline{)3\ 6\ ^6 4^4 5} \end{array}$$

We then repeat the process for 45:

$$1 \times 15 = 15$$
$$2 \times 15 = 30$$
$$3 \times 15 = 45 \qquad \textit{15 will go exactly 3 times}$$

The remainder will be zero. There are no more figures to consider, so the calculation is finished.

$$\begin{array}{r} 2\ 4\ 3 \\ 15\ \overline{)3\ 6\ ^6 4^4 5} \end{array}$$

So $3645 \div 15 = 243$.

The steps described can be written out like this:

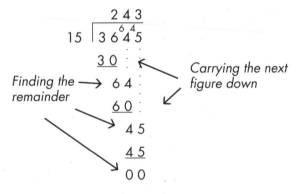

Finding the → remainder

Carrying the next figure down

This way of writing out all the steps is sometimes called long division.

Don't worry too much if you find the process complicated. You can always use a calculator to do division calculations like this.

Practice 9

1.	$165 \div 15$	6.	$8000 \div 20$
2.	$204 \div 12$	7.	$759 \div 11$
3.	$1536 \div 16$	8.	$2156 \div 22$
4.	$325 \div 25$	9.	$1800 \div 50$
5.	$2214 \div 18$	10.	$9225 \div 75$

Approximation

If we want an **approximate** (rough) answer to a calculation, we can round off the numbers in our calculation before we start.

Example

Donna is organizing a visit to a concert for 19 people. Each concert ticket costs £12, and Donna wants to work out roughly how much the total cost will be.

As she wants an approximate answer, Donna rounds the 19 up to 20, the £12 down to £10 and does the following calculation:

$$£10 \times 20 = £200$$

She has worked out that it is going to cost the group approximately £200.

Compare this with the accurate answer:

$$£12 \times 19 = £228$$

Donna has **estimated** the cost of the concert tickets.

Care must be taken when using approximations to find estimates. They should not be used when an accurate answer is required, but they can be used to give a rough idea of the size of the answer.

Approximations are useful for checking if answers found using a calculator are correct.

Example

Omar uses his calculator to multiply 5304 by 13. The answer he gets is 6942. He wants to carry out a rough check to see if his answer is right.

To do this, he rounds off the 5304 to 5000 and 13 to 10. He then does the calculation:

$$5000 \times 10 = 50\,000$$

This estimate tells him that his answer should be approximately fifty thousand, whereas the answer he got – 6942 – was much smaller than this. It looks like he made a mistake when entering the numbers in his calculator, so he goes back and checks the calculation again ... This time he gets 68 952, much closer to the estimate.

(Can you work out what mistake he made when he first entered the numbers?)

The following example shows another situation when using an estimate can be useful:

Example

Laminate flooring costs £8.50 for each square metre. John hasn't measured the floor of his living room, but he knows that it is approximately 4 metres wide by 5 metres long. Roughly how much will John have to spend on new flooring?

Approximate area of John's living room = 4m × 5m = 20 square metres

Approximate cost of flooring = £8.50 × 20 = £170

John will have to spend roughly £170.

It is useful to round off to the nearest 10, 100, 1000, and so on. This makes the calculation easier.

When rounding off to the nearest 10, a figure of 5 or more gets rounded up to the next 10 and a figure less than 5 gets rounded down.

13 gets rounded down to 10

16 gets rounded up to 20

When rounding off to the nearest 100, a figure of 50 or more gets rounded up to the next 100 and a figure less than 50 gets rounded down.

245 gets rounded down to 200

251 gets rounded up to 300

When rounding off to the nearest 1000, a figure of 500 or more gets rounded up to the next 1000 and a figure less than 500 gets rounded down.

2346 gets rounded down to 2000

2675 gets rounded up to 3000

Practice 10

Round off to the nearest 10:

1. 76
2. 34
3. 81
4. 279
5. 1237

Round off to the nearest 100:

6. 320
7. 550
8. 1379
9. 12501
10. 999

Negative numbers

All the numbers that we have considered so far have been positive whole numbers: +5, +3, +10, etc. Usually, there is no need to show the + sign in front of the number. We assume that any number without a sign in front of it is positive.

In number work we sometimes have to use negative numbers. These are numbers that are less than zero. They have a – sign in front of them: –7, –45, –23 and so on.

This might seem like a strange idea, but negative numbers can be useful in real life. Here are two situations that use negative numbers:

You have £300 in your bank account. You take out £350 to pay for a new computer, which means you have overdrawn your account by £50. You now have –£50 in your account.

BANK STATEMENT				
Date	Transactions	Credits	Debits	Balance
26 Sept	Supermarket		£10.59	£129.48
27 Sept	Direct Debit		£60.00	£69.48
28 Sept	Bookshop		£49.50	£19.98
28 Sept	Cheque Credit	£800.00		£819.98
29 Sept	Crazy Clothes		£6.99	£812.99
02 Oct	Delicatessen		£12.99	£800.00
02 Oct	Transfer		£500.00	£300.00
03 Oct	Crafty Computers		£350.00	−£50.00

One night last winter the outside temperature was 1° C. The following night it was 5 degrees lower. It fell to −4° C. (For more on temperature, see page 201.)

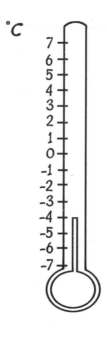

Prime numbers

A prime number is a number that can only be divided by 1 or itself. It is sometimes useful to be able to recognize prime numbers. For example, if you are finding the lowest form of a fraction, you won't be able to cancel down a prime number. (See page 49 in the section about fractions for more details.) Here are the prime numbers up to 100:

2, 3, 5, 7, 11, 13, 17, 19, 23, 29, 31, 37, 41, 43, 47, 53, 59, 61, 67, 71, 73, 79, 83, 89, 97

The order of operations

If a calculation has a mixture of +, −, × and ÷ within it, then there is a strict order in which the different parts of the calculation have to be done.

If you are working with whole numbers you have to carry out any multiplication or division *before* addition or subtraction.

Example

Calculate 3 + 5 × 4 − 8 ÷ 2.

When working this out we must do the multiplying and dividing first:

$$3 + 5 \times 4 - 8 \div 2$$

$$3 + 20 - 4$$

Now the + and − :

$$23 - 4 = 19$$

The correct answer is 19.

If the order of operations were ignored and the calculation worked from left to right (which might seem the obvious thing to do) the answer would have been 12, which would have been wrong.

 Most scientific calculators will automatically carry out the correct order of operations, so entering the calculation from left to right, ignoring the correct order, will produce the correct answer. However, most numeracy calculators will not.

Because this order of operations is not familiar to them, people often find it difficult to appreciate its importance. A practical example will show just how important it is:

> **Example**
>
> A plumber charges £50 call-out fee and £20 an hour for work carried out. If you employed this plumber, and the job took 3 hours to complete, how much would the work cost?

The calculation you would have to make to find out the cost would be:

£50 call-out fee + £20 every hour for 3 hours.

We can write this calculation like this:

$$£50 + 3 \times £20$$

Ignoring the rule of the order of operations and calculating from left to right, the total cost would be:

$$50 + 3 \ = \ 53$$
$$53 \times 20 = \ 1060$$

The cost of the job would be a whopping £1060!

Calculating correctly, doing the multiplying first, the total cost would be:

$$£50 \ + \ 3 \times £20 \ = \ £50 + £60 \ = \ £110$$

Which is a lot cheaper!

Sometimes people write brackets around the parts of the calculation that must be done first:

$$£50 + (3 \times £20)$$

 Some calculators have brackets that you can use to make sure that the operations are in the right order.

Practice 11

In this exercise try the questions first without using a calculator. If you have a calculator you could check your answers and at the same time check whether your calculator will automatically do the ordering for you.

1. $2 + 3 \times 4$

2. $8 - 6 \div 2$

3. $5 + 10 \div 5 - 2$

4. $13 - 3 \times 2 - 6 \div 2$

5. $4 \times 3 - 8 \div 4$

Problem solving

Number calculations often seem more difficult if presented as a written question. The words often obscure the calculation that has to be done.

Example

John is twelve years old. His father is four times older than him. John's mother is three years younger than her husband. How old is John's mother?

When tackling questions like this it is important not to rush at it and do the first thing that comes to mind. Read the question carefully and try to identify those parts of the question that relate to the calculation you need to carry out. Use a highlighter pen to try to identify these parts of the question.

Here is the problem again with the important parts highlighted.

John is **twelve years old**. His father is **four times older** than him. John's mother is **three years younger** than her husband. How old is John's mother?

Now write out the important information more simply:

John	12 years old
Father	4 times older
Mother	3 years less than father

Identify key words – **times, less** – and write the problem using symbols, in this case × and –.

John	12
Father	× 4
Mother	− 3

Finally put the problem together as a calculation:

| 12 × 4 | = | 48 |
| 48 − 3 | = | 45 |

John's mother is 45 years old.

This might seem a long process, and it isn't something you will need to do every time. Try using it, or your own version of it, if you get stuck on a written question because you can't understand what the question is asking you to do.

Practice 12

1. An adult numeracy class had sixteen students one week. The following week there were seven students fewer. How many students were in the class in the second week?

2. When Ahmed was twelve, his pocket money was £15 a month. On his 13th birthday his pocket money was increased by £4 a month. What was his new monthly pocket money?

3. The membership fee for Karina's swimming club is £32 a year. She also has to pay £3 each time she swims. In her first year of membership Karina went swimming 18 times. What was her total cost that year?

4. Joan's bus fare into town was 80p. She went into town and back three times in one week. What was her total bus fare for that week?

5. Michael earns £24 000 a year. How much is his monthly salary? (1 year = 12 months.)

FRACTIONS

What are fractions?

Fractions are used to describe parts of a whole.

If a pizza is shared equally between two people, each person gets half a pizza. A **half** is an example of a fraction, and is written as $\frac{1}{2}$.

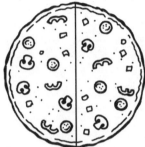

Fractions look very different from whole numbers, and this sometimes makes using them seem more difficult. You may be wondering why there are two numbers, and why one number is on top of the other.

The fraction $\frac{1}{2}$ is written in this way because it illustrates what has happened: 1 pizza has been divided into 2 equal pieces.

$$1 \div 2 \quad \rightarrow \quad \frac{1}{\div 2} \quad \rightarrow \quad \frac{1}{2}$$

Putting the numbers on top of each other is a way of showing that the top number has been divided by the bottom number.

If the pizza is shared between three people, each person would get a **third** of the pizza. A third is written as $\frac{1}{3}$.

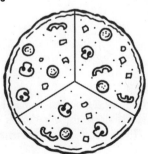

$$1 \div 3 \quad \rightarrow \quad \frac{1}{\div 3} \quad \rightarrow \quad \frac{1}{3}$$

Similarly, if the pizza is divided between four people, each person would get a **quarter**, which is written as $\frac{1}{4}$.

Now imagine the pizza divided into four equal pieces ...

... and someone eats one of the pieces, leaving only three pieces.

Three quarters are left.

This is written as $\frac{3}{4}$. ← *The top number of the fraction tells us how many quarters are left.*

The bottom number tells us how many equal pieces the pizza was first divided into.

 The **bottom** number of a fraction shows how many equal pieces the whole has been divided into. It shows what **type** of fraction we are dealing with. This is called the **denominator**.

The **top** number of a fraction shows **how many** of the pieces we are using. This is called the **numerator**.

Understanding the different roles of the two numbers of a fraction will help you in your fraction work.

Naming fractions

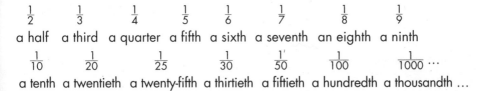

$\frac{1}{2}$ $\frac{1}{3}$ $\frac{1}{4}$ $\frac{1}{5}$ $\frac{1}{6}$ $\frac{1}{7}$ $\frac{1}{8}$ $\frac{1}{9}$

a half   a third   a quarter   a fifth   a sixth   a seventh   an eighth   a ninth

$\frac{1}{10}$ $\frac{1}{20}$ $\frac{1}{25}$ $\frac{1}{30}$ $\frac{1}{50}$ $\frac{1}{100}$ $\frac{1}{1000}$ $\cdots$

a tenth   a twentieth   a twenty-fifth   a thirtieth   a fiftieth   a hundredth   a thousandth ...

Most fractions are named by adding '-th' on to the end of the bottom number. Sometimes the spelling of the number changes slightly; for example 'a ninth' rather than 'a nineth' and 'a twentieth' rather than 'a twentyth'. This makes them easier to say.

You will notice that the first three fractions in the list are named in a different way from the rest. This is probably because it would be difficult to say 'a twoth', or 'a threeth', though sometimes a quarter is called 'a fourth'.

There are just a couple more exceptions, for fractions with a bottom number ending in 1 or 2:

$\frac{1}{31}$ $\qquad\qquad\qquad$ $\frac{1}{62}$

a thirty-first $\qquad\qquad$ a sixty-second

Equivalent fractions

Another important idea in fraction work is that a fraction can be written in many different ways and yet still represent exactly the same amount. This is called **equivalence**.

Consider the fraction $\frac{1}{2}$ by thinking about half of a pizza.

Fractions

Pizza divided between two people   Pizza divided between four people

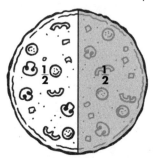   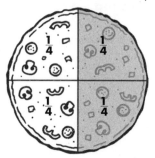

You can see from the two pizzas that the quantity for a half is the same as the quantity for two quarters.

$$\frac{1}{2} = \frac{2}{4}$$

The fraction $\frac{2}{4}$ looks different from $\frac{1}{2}$ but it represents the same amount.

We can extend this idea by imagining the pizza divided between 8 people.

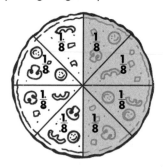

From the diagram, we can see that $\frac{4}{8}$ of the pizza is the same as $\frac{2}{4}$ and $\frac{1}{2}$.

$$\frac{4}{8} = \frac{2}{4} = \frac{1}{2}$$

Again, the fractions look very different, but they all represent the same amount of pizza. We can extend this idea further with the pizza being shared between 16, 32, 64 people and so on.

$$\frac{1}{2} = \frac{2}{4} = \frac{4}{8} = \frac{8}{16} = \frac{16}{32} = \frac{32}{64}$$

All these fractions represent half a pizza.

Similarly, if we consider the pizza divided between 3 people, we can show that:

$$\frac{1}{3} = \frac{2}{6} = \frac{4}{12} = \frac{8}{24} = \frac{16}{48}$$

All these fractions still represent one third of a pizza.

This idea can be applied to *any* fraction. For example,

$$\frac{1}{5} = \frac{2}{10} = \frac{4}{20} = \frac{8}{40} = \quad \dots$$

The ability to write fractions in different forms is a key concept in fraction work, and it is worth making sure you understand it before moving on. Later in the fraction section, we will be using the idea to add and subtract fractions.

Drawing pizzas is a useful way of illustrating the idea of equivalence, but it would be more convenient to be able to write equivalent fractions without having to draw pizzas.

Have a look at this series of fractions and try to work out what is happening to the top and bottom numbers each time.

$$\frac{1}{2} = \frac{2}{4} = \frac{4}{8} = \frac{8}{16} = \frac{16}{32} = \frac{32}{64}$$

What seems to be happening to the top and bottom numbers of the fraction?

The top and bottom numbers have been doubled each time.

$$\frac{1 \times 2}{2 \times 2} = \frac{2 \times 2}{4 \times 2} = \frac{4 \times 2}{8 \times 2} = \frac{8 \times 2}{16 \times 2} = \frac{16 \times 2}{32 \times 2} = \frac{32}{64}$$

This gives us a quick and convenient way of creating equivalent fractions.

 Multiplying the top and bottom number of a fraction by the same whole number creates an equivalent fraction.

In the examples so far the top and bottom numbers have been multiplied by 2, but it can be *any* whole number providing you use the *same* number to multiply top and bottom.

Example

How many tenths ($\frac{1}{10}$) are equivalent to one half?

We can write this as:

$$\frac{1}{2} \quad = \quad \frac{?}{10}$$

The 2 has been changed to 10 by multiplying by 5, so the top number – the 1 – must also be multiplied by 5. $1 \times 5 = 5$, so the missing number is 5.

$$\frac{1}{2} \xrightarrow[\times 5]{=} \frac{?}{10} \qquad\qquad\qquad \frac{1}{2} \xrightarrow[]{\times 5} = \frac{5}{10}$$

Five tenths are equivalent to one half.

Practice 1

In the following exercise, one of the numbers of each fraction has been changed. See if you can work out what that number has been multiplied by, and multiply the other number accordingly.

1. $\frac{1}{2} = \frac{?}{14}$ 6. $\frac{1}{7} = \frac{?}{70}$

2. $\frac{1}{4} = \frac{?}{12}$ 7. $\frac{3}{4} = \frac{?}{20}$

3. $\frac{1}{3} = \frac{3}{?}$ 8. $\frac{2}{5} = \frac{8}{?}$

4. $\frac{1}{6} = \frac{?}{18}$ 9. $\frac{5}{9} = \frac{?}{27}$

5. $\frac{1}{8} = \frac{3}{?}$ 10. $\frac{2}{7} = \frac{10}{?}$

 If you can't 'see' what the top or bottom number has been multiplied by, you can use a multiplying square as shown on page 19 in the section about multiplying whole numbers.

Cancelling

The process of creating equivalent fractions by multiplying up can be reversed. Often equivalent fractions can be created by dividing the top and bottom number by a whole number. This is called **cancelling** or **cancelling down**.

When multiplying up, we saw that you could choose to multiply by *any* whole number, as long as you used the *same* whole number to multiply top and bottom of the fraction.

When cancelling, you can divide by any number as long as that number will divide into both the top and bottom numbers of the fraction to give whole number answers (otherwise you will end up with bits left over at the top or bottom).

$$\frac{4 \div 2}{6 \div 2} = \frac{2}{3} \checkmark \qquad \frac{18 \div 3}{24 \div 3} = \frac{6}{8} \checkmark \qquad \frac{60 \div 20}{100 \div 20} = \frac{3}{5} \checkmark$$

But

$$\frac{4 \div 3}{6 \div 3} = \frac{1\,R1}{3} \; \times$$

Notice that $\frac{18}{24}$ could also be cancelled by 2 or 6 instead of 3:

$$\frac{18 \div 2}{24 \div 2} = \frac{9}{12} \qquad \frac{18 \div 6}{24 \div 6} = \frac{3}{4}$$

In the same way, $\frac{60}{100}$ could be cancelled by 2, 5 or 10.

Cancelling to lowest or simplest form

We saw earlier that the fraction 'a half' can be written in many different forms:

$$\frac{1}{2} = \frac{2}{4} = \frac{4}{8} = \frac{8}{16} = \frac{16}{32} = \frac{32}{64}$$

We also saw earlier that these different forms can be made by multiplying or dividing the top and bottom numbers.

$\frac{1}{2}$ is the **lowest form** of these fractions because the numbers 1 and 2 cannot both be divided down further by any whole number to give whole number answers.

 To cancel a fraction to its lowest form, divide the top and bottom number of the fraction repeatedly until you can't divide further.

Remember that, having cancelled the fraction once by one number, you can sometimes cancel again using either the same number or a different number. The skill is finding numbers that will divide exactly into both top and bottom numbers of the fraction. This gets easier with practice, but the multiplying square can be helpful as shown in the following example.

Cancel $\frac{16}{24}$ to its lowest form.

Have a look at the multiplying square below.

We want to find a number that will divide into 16 and 24. A column or row with both 16 and 24 in it will show you a number that will divide into both.

	1	2	3	4	5	6	7	8	9	10
1	1	2	3	4	5	6	7	8	9	10
2	2	4	6	8	10	12	14	16	18	20
3	3	6	9	12	15	18	21	24	27	30
4	4	8	12	16	20	24	28	32	36	40
5	5	10	15	20	25	30	35	40	45	50
6	6	12	18	24	30	36	42	48	54	60
7	7	14	21	28	35	42	49	56	63	70
8	8	16	24	32	40	48	56	64	72	80
9	9	18	27	36	45	54	63	72	81	90
10	10	20	30	40	50	60	70	80	90	100

Following the arrows shows that 4 will divide into both 16 and 24.

Dividing top and bottom numbers by 4:

$$\frac{16 \div 4}{24 \div 4} = \frac{4}{6}$$

Repeat the process by trying to find a row or column in the multiplying square which contains both a 4 and a 6.

4 and 6 can both be divided by 2:

$$\frac{4 \div 2}{6 \div 2} = \frac{2}{3}$$

You might notice that in the multiplying square 4 and 6 both appear in the row and the column of the 1s. We can ignore this, since dividing by 1 doesn't change a number.

Repeat the process by trying to find a row or column in the multiplying square which contains both a 2 and a 3. Since there is no row or column that has both a 2 and a 3 in it (except the row and the column of the 1s which we can ignore), we can conclude that there is no whole number that will divide into both 2 and 3.

$\frac{2}{3}$ is therefore the lowest form of the fraction.

100 is the largest number in the multiplying square. If a fraction contains numbers greater than 100, we can't use the number square to find a number that will divide. Instead, you can use a 'trial and error' method:

Try dividing the top and bottom number of the fraction by 2. If 2 doesn't divide into both exactly, then try 3, if 3 doesn't divide exactly then try 5. If 5 doesn't divide into both exactly, then try 7, 11, 13, 17 or one of the other prime numbers (for a reminder about prime numbers, see page 37 in the section about whole numbers).

It's also worth remembering that if both numbers end in zero, they can both be divided by 10. If the numbers end in 0 or 5 they can be divided by 5.

Some fractions are already in their lowest forms. They cannot be cancelled down further.

Practice 2

Cancel the following fractions to their lowest forms:

1. $\frac{12}{15}$ 6. $\frac{36}{60}$

2. $\frac{40}{60}$ 7. $\frac{20}{45}$

3. $\frac{18}{30}$ 8. $\frac{36}{45}$

4. $\frac{16}{30}$ 9. $\frac{120}{400}$

5. $\frac{21}{56}$ 10. $\frac{246}{300}$

Finding a fraction of a number

The ability to find a fraction of a number is a skill that is useful in many everyday situations.

Example

A sweater originally cost £48. If it is marked '$\frac{2}{3}$ off', what does it cost now?

To find the sale price of the article, we would have to find $\frac{2}{3}$ of £48 and deduct it from the normal price.

We will come back to this example shortly but, before we do, try answering the following questions:

'What is a half ($\frac{1}{2}$) of £10?' (If it helps, imagine sharing £10 between two people.)

'What is a third ($\frac{1}{3}$) of £12?' (If it helps, imagine sharing £12 between three people.)

'What is a quarter ($\frac{1}{4}$) of £8?' (If it helps, imagine sharing £8 between four people.)

You probably worked out the answers to be £5, £4, and £2.

Now ask yourself the question: 'How did I calculate these answers?'

Very often, the answers to questions like these seem just to pop into our heads. In the past we have had to calculate a half, or a third, or a quarter of an amount enough times to make it seem an automatic process. What we are actually doing is:

dividing by 2 to find a half:	£10 ÷ 2 = £5
dividing by 3 to find a third:	£12 ÷ 3 = £4
dividing by 4 to find a quarter:	£8 ÷ 4 = £2

This gives us a method for finding any **unitary** fraction of a number – that is, any fraction with a top number of 1.

 To find a unitary fraction of an amount, divide the amount by the bottom number of the fraction.

For example:

$\frac{1}{5}$ of £20 = £20 ÷ 5 = £4

$\frac{1}{6}$ of £18 = £18 ÷ 6 = £3

$\frac{1}{50}$ of £200 (£200 shared between 50 people) = £200 ÷ 50 = £4

Now back to the example of buying a sweater. In the sale, the initial price (£48) was reduced by $\frac{1}{3}$. To find the reduction we need to calculate $\frac{2}{3}$ of the normal price.

$\frac{2}{3}$ of £48 = ?

First, find $\frac{1}{3}$ by dividing by 3:

£48 ÷ 3 = £16

If $\frac{1}{3}$ (one third) is £16, then $\frac{2}{3}$ (two thirds) will be:

£16 × 2 = £32

So there will be a reduction of £32 in the price of the sweater. The sale price will be:

£48 – £32 = £16

This gives a general rule for finding any fraction of a number:

 To find a fraction of an amount, divide the amount by the bottom number of the fraction and multiply the answer by the top number of the fraction.

Fractions

Practice 3

Find the following fractions:

1. $\frac{1}{5}$ of £225
2. $\frac{1}{3}$ of £90
3. $\frac{3}{4}$ of £720
4. $\frac{4}{5}$ of £25
5. $\frac{7}{9}$ of £81

6. $\frac{5}{6}$ of £36
7. $\frac{9}{10}$ of £100
8. $\frac{7}{9}$ of £18
9. $\frac{3}{7}$ of £21
10. $\frac{5}{5}$ of £25

Writing one number as a fraction of another

Sometimes it is necessary to write one number as a fraction of another, as in the following example:

Example

Out of 24 candidates for a job, 8 are women. What fraction of the candidates are women?

8 out of 24 are women. This can be expressed as a fraction by putting the 8 over the 24.

$\frac{8}{24}$

The fraction should then be cancelled down if possible.

$\frac{\cancel{8}}{\cancel{24}} \ = \ \frac{1}{3}$ (dividing top and bottom by 8)

A third $(\frac{1}{3})$ of the group are women.

Practice 4

Write the first number as a fraction of the second in each question. You should cancel down to the lowest term wherever possible.

1.   2 as a fraction of 3
2.   3 as a fraction of 12
3.   18 as a fraction of 20
4.   15 as a fraction of 22
5.   36 as a fraction of 40

6.   9 as a fraction of 18
7.   27 as a fraction of 90
8.   100 as a fraction of 450
9.   15 as a fraction of 100
10. 21 as a fraction of 84

Top-heavy and mixed fractions

The following diagram shows one and a half pizzas:

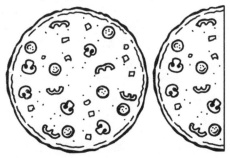

This can be written as $1\frac{1}{2}$.

Alternatively, one and a half pizzas could be thought of as three halves of pizza:

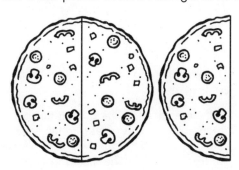

This can be written as $\frac{3}{2}$.

$$1\frac{1}{2} = \frac{3}{2}$$

$1\frac{1}{2}$ is an example of a **mixed fraction** – it is a mixture of a whole number and a fraction.

$\frac{3}{2}$ is an example of a **top-heavy fraction** – top-heavy because the top number is bigger than the bottom number. A top-heavy fraction can also be called an **improper** fraction.

Mixed fractions can be written as top-heavy fractions and vice versa.

Changing mixed fractions into top-heavy fractions

A mixed fraction can be changed to the equivalent top-heavy fraction by changing the whole number part of the mixed fraction into a fraction and adding to the fraction part.

Example

Change $2\frac{3}{4}$ into a top-heavy fraction.

First, change the 2 into quarters. There are 4 quarters in one whole, so in 2 there will be eight quarters.

$$2 = \frac{8}{4}$$

Add to the fraction part:

$$\frac{8}{4} + \frac{3}{4} = \frac{11}{4} \quad \text{(we will look more at adding fractions later in this section)}$$

So $2\frac{3}{4} = \frac{11}{4}$

Here is a quick method for converting a mixed fraction to a top-heavy fraction:

 Multiply the whole number by the bottom number of the fraction and add the top number. This gives you the top number of the top-heavy fraction. The bottom number will be the same as the bottom number of the fraction part of the mixed fraction.

Example

What is $2\frac{3}{4}$ as a top-heavy fraction?

$$2 \times 4 = 8$$
$$8 + 3 = 11$$
$$2\tfrac{3}{4} = \tfrac{11}{4}$$

Practice 5

Change the following mixed fractions into top-heavy fractions:

1.	$1\tfrac{3}{8}$		6.	$5\tfrac{7}{8}$
2.	$3\tfrac{5}{8}$		7.	$10\tfrac{1}{2}$
3.	$5\tfrac{1}{2}$		8.	$13\tfrac{1}{2}$
4.	$6\tfrac{2}{3}$		9.	$12\tfrac{2}{3}$
5.	$2\tfrac{3}{4}$		10.	$50\tfrac{1}{4}$

Changing top-heavy fractions into mixed fractions

A top-heavy fraction can be changed into a mixed fraction by calculating how many whole numbers can be made from the top-heavy fraction.

Example

Change $\tfrac{9}{4}$ into a mixed fraction.

Four quarters make one whole, so eight quarters make two wholes. This leaves one spare quarter.

$$\tfrac{9}{4} = \tfrac{8}{4} + \tfrac{1}{4} = 2 + \tfrac{1}{4} = 2\tfrac{1}{4}$$

Here is a quick method for converting a top-heavy fraction to a mixed fraction:

 Divide the bottom number into the top number. The answer is the whole number of the mixed fraction; the remainder is the fraction.

Example

What is $\frac{5}{3}$ as a mixed fraction?

Dividing 3 into 5:

$$3\overline{)5}^{1\,R\,2}$$

The equivalent mixed fraction is $1\frac{2}{3}$. Notice that the bottom number of the fraction part of the mixed fraction is the same as the bottom number of the top-heavy fraction.

Practice 6

Change the following top-heavy fractions into mixed fractions:

1. $\frac{8}{3}$

2. $\frac{9}{2}$

3. $\frac{10}{3}$

4. $\frac{15}{7}$

5. $\frac{17}{5}$

6. $\frac{23}{7}$

7. $\frac{50}{9}$

8. $\frac{13}{12}$

9. $\frac{120}{7}$

10. $\frac{1001}{200}$

Top-heavy fractions and whole numbers

Sometimes converting a top-heavy fraction gives a whole number answer, so there is no fraction part.

$$\frac{20}{5} = 4$$

Whenever the top number of a fraction is the same as the bottom number it is equal to one.

$$\frac{5}{5} = 1$$

Ordering fractions

Consider these fractions: $\frac{1}{4}, \frac{5}{4}, \frac{3}{4}$.

It is fairly easy to write these fractions in order of size, biggest first, because they are all quarters. $\frac{5}{4}$ is the biggest, then $\frac{3}{4}$ and finally $\frac{1}{4}$.

Now consider these fractions: $\frac{2}{5}, \frac{7}{10}, \frac{12}{20}$.

It is harder to put these fractions in order, biggest first, because they all have different denominators (bottom numbers). However, by using the idea of equivalent fractions we can make them all into the same type of fraction.

$$\frac{2}{5} \times \frac{4}{4} = \frac{8}{20}$$

$$\frac{7}{10} \times \frac{2}{2} = \frac{14}{20}$$

The fractions now look like this: $\frac{8}{20}, \frac{14}{20}, \frac{12}{20}$.

Now it is easier to put them in order. The biggest is $\frac{14}{20}$, the next biggest is $\frac{12}{20}$ and the smallest is $\frac{8}{20}$.

Using the original fractions, the biggest is $\frac{7}{10}$, the next biggest is $\frac{12}{20}$ and the smallest is $\frac{2}{5}$.

Practice 7

Put the following fractions in order, biggest first:

1. $\frac{2}{3}, \frac{3}{4}, \frac{1}{2}$
2. $\frac{3}{5}, \frac{7}{10}, \frac{1}{2}$
3. $\frac{2}{7}, \frac{3}{5}, \frac{12}{35}$
4. $\frac{2}{9}, \frac{1}{10}, \frac{3}{5}$
5. $\frac{2}{3}, \frac{7}{8}, \frac{5}{12}$

Adding fractions

Although we don't often have to add fractions in everyday life, there are occasions when this skill is useful, as in the following example:

Example

Gordon arrived at the station half an hour before his train was due. The person in the ticket office told him that the train was running a quarter of an hour late. How long would Gordon have to wait before his train came?

To find Gordon's total waiting time we need to add together $\frac{1}{2}$ an hour and $\frac{1}{4}$ of an hour.

$$\frac{1}{2} + \frac{1}{4}$$

You might know the answer to this without having to do the calculation, because adding together quantities of time is something that we do quite often. However, it will be useful to work through the calculation, as this will show us a method we can use to add fractions in other contexts.

Look at the half hour and the quarter of an hour on the faces of two clocks:

The half hour can be broken down into two quarters:

So the question becomes 'two quarters plus one quarter', which makes three quarters.

$$\frac{2}{4} + \frac{1}{4} = \frac{3}{4}$$

Gordon would have to wait three quarters ($\frac{3}{4}$) of an hour for his train.

Here are some general tips for adding fractions:

 Fractions can only be added if the bottom numbers (the denominators) are the same.

If the bottom numbers are the same, addition is carried out by adding **the top numbers only**. The bottom number (the denominator) of the answer will be the same as the bottom numbers of the fractions we are adding.

 If the bottom numbers of the fractions to be added are not the same, we have to make them the same before adding.

Remember, that in any fraction, the bottom number simply tells us the *type* of fraction we are working with – halves, quarters, thirds, and so on. It is the top number that tells us how *many* of that fraction we have.

Making the bottom numbers the same

 This is also called **finding a common denominator**.

To make the bottom numbers the same, we use the idea of equivalent fractions (looked at earlier in this section), to change one or more of the fractions. Two examples will show how this is done.

Example

What is $\frac{1}{4} + \frac{3}{8}$?

The first thing to check is whether the bottom numbers are the same. If they are, you can simply add the top numbers, keeping the same bottom number, as shown in an earlier example.

If, as in this example, the bottom numbers are different, check whether one of the bottom numbers can be multiplied up to equal the other number.

In this example, 4 can be multiplied by 2 to make 8.

$$\frac{1}{4} \quad \begin{matrix} \times \\ \times \end{matrix} \quad \begin{matrix} 2 \\ 2 \end{matrix} \quad = \quad \frac{2}{8}$$

 Notice that, having decided to multiply the 4 by 2 to turn it into 8, we have to multiply the top number of the fraction by 2 as well. Otherwise, the fraction we are creating is not equivalent to $\frac{1}{4}$.

So instead of $\frac{1}{4}$ we can write $\frac{2}{8}$.

$$\frac{1}{4} \quad + \quad \frac{3}{8} \quad = \quad \frac{2}{8} \quad + \quad \frac{3}{8}$$

Since the bottom numbers are now the same we can add the two fractions:

$$\frac{2}{8} \quad + \quad \frac{3}{8} \quad = \quad \frac{5}{8}$$

Example

What is $\frac{1}{2} + \frac{1}{3}$?

Check: Are the bottom numbers the same? **No.**

Check: can one number be multiplied (by a whole number) to make it the same as the other? **No.**

So, in this example, both numbers have to be changed into another common number.

Look for a number that *both* 2 and 3 can be multiplied up to. Or, alternatively, look for a number that 2 and 3 will both divide into giving whole number answers.

In this example we could use 6, 12, 18, 24, ... the list goes on for ever. These are all **multiples** of 2 and 3 ($2 \times 3 = 6$, $6 \times 2 = 12$, $6 \times 3 = 18$, $12 \times 2 = 24$, and so on). We could use any one of these numbers, but using the smallest is easiest because it will leave us with less cancelling to do at the end of the calculation. The smallest number is called the **lowest common denominator**.

Each fraction now has to be changed into an equivalent fraction with the new denominator (bottom number).

$$\frac{1}{2} \quad = \quad \frac{?}{6} \qquad\qquad \xrightarrow{\times 3} \qquad \frac{1}{2} \quad = \quad \frac{3}{6}$$
$$\xrightarrow{\times 3}$$

The 2 has been multiplied by 3 to change it into 6, so the 1 must also be multiplied by 3.

$$\frac{1}{3} \quad = \quad \frac{?}{6} \qquad\qquad \xrightarrow{\times 2} \qquad \frac{1}{3} \quad = \quad \frac{2}{6}$$
$$\xrightarrow{\times 2}$$

The 3 has been multiplied by 2 to change it into 6, so the 1 must also be multiplied by 2.

Having made these changes, the addition becomes:

$$\frac{3}{6} \quad + \quad \frac{2}{6}$$

Since the bottom numbers are now the same we can add the top numbers together.

$$\frac{3}{6} \quad + \quad \frac{2}{6} \quad = \quad \frac{5}{6}$$

Practice 8

Add the following fractions and give your answer as a fraction in its lowest form:

1. $\frac{1}{3} + \frac{1}{3}$

2. $\frac{1}{5} + \frac{2}{5}$

3. $\frac{1}{3} + \frac{1}{6}$

4. $\frac{3}{4} + \frac{1}{12}$

5. $\frac{5}{7} + \frac{4}{21}$

6. $\frac{2}{3} + \frac{3}{5}$

7. $\frac{5}{6} + \frac{3}{4}$

8. $\frac{1}{3} + \frac{5}{6}$

9. $\frac{8}{10} + \frac{3}{4}$

10. $\frac{2}{5} + \frac{3}{4} + \frac{1}{3}$

 The lowest common denominator is sometimes called the **lowest common multiple** (LCM).

Adding mixed fractions

There are two methods of adding mixed fractions.

Method 1: change the mixed fractions to top-heavy (improper) fractions and add as above.

Example

Add $2\frac{1}{2}$ to $3\frac{3}{4}$.

Fractions

$$2\frac{1}{2} + 3\frac{3}{4} \quad = \quad \frac{5}{2} + \frac{15}{4} \quad = \quad \frac{10}{4} + \frac{15}{4} \quad = \quad \frac{25}{4} \quad = \quad 6\frac{1}{4}$$

Method 2: add the whole number and fraction parts separately. Using the same example:

Example

Add $2\frac{1}{2}$ to $3\frac{3}{4}$.

Add the whole numbers:

$2 + 3 = 5$

Add the fractions:

$$\frac{1}{2} + \frac{3}{4} \quad = \quad \frac{2}{4} + \frac{3}{4} \quad = \quad \frac{5}{4} \quad = \quad 1\frac{1}{4}$$

Add the two parts of the calculation together:

$$5 + 1\frac{1}{4} \quad = \quad 6\frac{1}{4}$$

Practice 9

Add the following mixed fractions. Try both methods and decide which one you prefer.

1. $2\frac{1}{8} + 1\frac{1}{4}$
2. $3\frac{3}{4} + 4\frac{1}{2}$
3. $4\frac{1}{2} + 5\frac{1}{3}$
4. $10\frac{3}{8} + 2\frac{2}{3}$
5. $6\frac{3}{4} + 1\frac{1}{3}$

Subtracting fractions

The subtraction of fractions can be done in the same way as addition. At the end of the calculation, the top numbers are subtracted from each other instead of being added together.

Example

Calculate $\frac{3}{5} - \frac{1}{2}$.

The lowest common denominator for the two fractions is 10, so we must change the bottom numbers of each fraction to 10. For $\frac{3}{5}$, we multiply up the top and bottom numbers by 2, and for $\frac{1}{2}$ we multiply up by 5.

$$\frac{3}{5} - \frac{1}{2} \quad = \quad \frac{6}{10} - \frac{5}{10} \quad = \quad \frac{1}{10}$$

Practice 10

1. $\frac{7}{9} - \frac{2}{9}$

2. $\frac{2}{3} - \frac{1}{6}$

3. $\frac{5}{10} - \frac{3}{20}$

4. $\frac{8}{9} - \frac{1}{18}$

5. $\frac{3}{4} - \frac{1}{5}$

6. $\frac{2}{3} - \frac{5}{10}$

7. $\frac{6}{7} - \frac{2}{5}$

8. $\frac{2}{3} - \frac{7}{20}$

9. $\frac{11}{12} - \frac{2}{5}$

10. $\frac{1}{8} - \frac{1}{9}$

Subtracting mixed fractions

The subtraction of mixed fractions can be done in the same way as addition. Just remember to subtract instead of adding! As with adding mixed fractions, there are two methods.

Method 1: change the mixed fractions to top-heavy (improper) fractions and subtract the top numbers.

Example

What is $2\frac{3}{4} - 1\frac{1}{2}$?

$$2\frac{3}{4} - 1\frac{1}{2} \quad = \quad \frac{11}{4} - \frac{3}{2} \quad = \quad \frac{11}{4} - \frac{6}{4} \quad = \quad \frac{5}{4} \quad = \quad 1\frac{1}{4}$$

Fractions

Method 2: subtract the whole number and fraction parts separately.

Example

What is $2\frac{3}{4} - 1\frac{1}{2}$?

First subtract the whole numbers:

$2 - 1 = 1$

Then subtract the fractions:

$$\frac{3}{4} - \frac{1}{2} \quad = \quad \frac{3}{4} - \frac{2}{4} \quad = \quad \frac{1}{4}$$

Add the two parts of the calculation together:

$$1 + \frac{1}{4} \quad = \quad 1\frac{1}{4}$$

A difficulty sometimes arises when subtracting mixed fractions using method 2, as shown by the following example.

Example

What is $2\frac{1}{2} - 1\frac{3}{4}$?

The number $2\frac{1}{2}$ is bigger than the number $1\frac{3}{4}$, so the calculation can be done. However, the fraction part of the first number ($\frac{1}{2}$) is smaller than the fraction part of the second number ($\frac{3}{4}$), so we can't subtract them (without entering the world of negative numbers!)

We can get around this problem by borrowing one from the 2 as shown below:

$$1\,\cancel{2}\frac{1}{2} \quad - \quad 1\frac{3}{4}$$

The 1 that we've borrowed can be changed into two halves ($\frac{2}{2}$) and added to the $\frac{1}{2}$ that is already there, so the calculation becomes

$$1\frac{3}{2} - 1\frac{3}{4}$$

Subtract the whole numbers:

$1 - 1 = 0$

Then subtract the fractions:

$\frac{3}{2} - \frac{3}{4}$ = $\frac{6}{4} - \frac{3}{4}$ = $\frac{3}{4}$

This method is quite complicated. In general, when subtracting mixed fractions it is better to convert both fractions to top-heavy fractions as we did in method 1.

Practice 11

1. $3\frac{1}{2} - 1\frac{3}{4}$
2. $2\frac{1}{4} - 1\frac{1}{2}$
3. $5\frac{1}{4} - 3\frac{3}{4}$
4. $7\frac{2}{3} - 2\frac{1}{8}$
5. $5\frac{5}{8} - 3\frac{1}{3}$

Multiplying fractions

Multiplying a fraction by a whole number

We saw in the section about whole numbers that multiplying whole numbers is a short-cut to repeated addition (see page 18).

$5 + 5 + 5 + 5 + 5 + 5 + 5$ (seven fives) $= 7 \times 5 = 35$

Similarly, multiplying a fraction by a whole number is a short-cut to repeated additon of that fraction.

$5 \times \frac{1}{2}$ = $\frac{1}{2} + \frac{1}{2} + \frac{1}{2} + \frac{1}{2} + \frac{1}{2}$ (five halves) = $\frac{5}{2}$ = $2\frac{1}{2}$

A quick method of doing this is by multiplying the *top* number of the fraction by the whole number.

$$5 \times \frac{1}{2} = \frac{5 \times 1}{2} = \frac{5}{2} = 2\frac{1}{2}$$

Practice 12

1. $4 \times \frac{3}{5}$
2. $8 \times \frac{5}{7}$
3. $2 \times \frac{3}{4}$
4. $10 \times \frac{1}{5}$
5. $7 \times \frac{4}{9}$

Multiplying a fraction by a fraction

Earlier in this section we looked at finding a fraction of a whole number. When multiplying one fraction by another we are, in effect, finding a fraction of another fraction.

Example

What is $\frac{1}{2} \times \frac{1}{4}$?

$\frac{1}{2} \times \frac{1}{4}$ means $\frac{1}{2}$ **of** $\frac{1}{4}$.

The first diagram shows $\frac{1}{4}$.

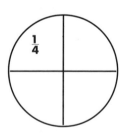

The next diagram shows that a half of a quarter is an eighth.

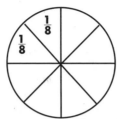

In practice, we can get the same answer by multiplying the top numbers and multiplying the bottom numbers.

$$\frac{1}{2} \times \frac{1}{4} = \frac{1 \times 1}{2 \times 4} = \frac{1}{8}$$

 To multiply two fractions multiply the top numbers **and** multiply the bottom numbers. Cancel your answer down if you can.

 In fraction calculations, **x** and **of** are interchangeable.

$\frac{1}{2}$ of $\frac{1}{4}$ is the same as $\frac{1}{2} \times \frac{1}{4}$.

$\frac{1}{4}$ of $\frac{3}{4}$ is the same as $\frac{1}{4} \times \frac{3}{4}$.

Multiplying more than two fractions

In the examples we've looked at we have multiplied two fractions. The same method of multiplying top numbers and multiplying bottom numbers can be used when more than two fractions are multiplied.

Example

What is $\frac{1}{2}$ times $\frac{3}{4}$ times $\frac{2}{3}$?

$$\frac{1}{2} \times \frac{3}{4} \times \frac{2}{3} = \frac{1 \times 3 \times 2}{2 \times 4 \times 3} = \frac{6}{24}$$

Cancelling after multiplying

When we multiply two (or more) fractions, the answer can sometimes be cancelled to a lower term.

Example

Calculate $\frac{1}{2} \times \frac{4}{5}$.

Fractions

$$\frac{1}{2} \times \frac{4}{5} = \frac{4}{10} = \frac{2}{5} \quad \text{(dividing top and bottom by 2)}$$

Cancelling before multiplying

Sometimes it is possible to cancel before multiplying the fractions.

Example

Calculate $\frac{3}{8} \times \frac{4}{9}$.

First, we look to see if we can cancel any number on the top with any number on the bottom. As in cancelling individual fractions, this means looking for a whole number that will divide into a number on the top of the multiplication and a number on the bottom. However, unlike cancelling individual fractions, the numbers on the top and bottom don't have to belong to the same fraction (see below).

In this example, 3 will divide into the 3 on the top and into the 9 on the bottom. 4 will divide into the 4 on the top and the 8 on the bottom.

$$\frac{^1\cancel{3}}{8} \times \frac{4}{\cancel{9}_3} = \frac{1}{_2\cancel{8}} \times \frac{\cancel{4}^1}{3} = \frac{1}{2} \times \frac{1}{3} = \frac{1}{6}$$

Don't worry – if you forget to cancel before multiplying, you can always cancel afterwards.

Multiplying mixed fractions

To multiply mixed fractions, change them to top-heavy fractions and then use the above method.

Practice 13

1. $\frac{1}{2} \times \frac{3}{4}$
2. $\frac{3}{5} \times \frac{5}{6}$
3. $\frac{4}{9} \times \frac{3}{8}$
4. $1\frac{1}{2} \times 2\frac{3}{4}$
5. $4\frac{3}{4} \times 5\frac{1}{3}$

Dividing fractions

Think about the following question: what is four divided by a half?

Many people will interpret this question as meaning 'find a half of four', which would give an answer of 2. In fact, the question is asking us to calculate *how many halves are in four*. The answer to this question is 8.

It is easy to make this mistake, so it is worth making sure you understand the difference between the two questions.

$$\frac{1}{2} \text{ of } 4 \quad = \quad \frac{1}{2} \times 4 \quad = \quad 2$$

Or, alternatively:

$$\frac{1}{2} \times 4 \quad = \quad 4 \times \frac{1}{2} \quad = \quad \text{four halves} \quad = \quad \frac{1}{2} + \frac{1}{2} + \frac{1}{2} + \frac{1}{2} \quad = \quad 2$$

but

$$4 \div \frac{1}{2} \quad =$$

There are 8 halves in 4. So $4 \div \frac{1}{2} = 8$.

Now think about what happens if, instead of dividing 4 by $\frac{1}{2}$, we multiply 4 by 2:

$$4 \times 2 = 8$$

We get the same answer.

Multiplication is the reverse of division, so multiplying by the reversed fraction (that is, the fraction turned 'upside down') gives the same answer.

Turning $\frac{1}{2}$ upside down gives $\frac{2}{1}$ or 2:

$$4 \div \frac{1}{2} \quad = \quad 4 \times \frac{2}{1} \quad = \quad 4 \times 2 \quad = \quad 8$$

This gives us an easy method of dividing by a fraction, which doesn't involve trying to picture how many times the fraction goes into the number we are dividing into.

 When dividing by a fraction, change division to multiplication but turn the fraction by which you are dividing 'upside down'.

Dividing a fraction by a fraction

In the previous example, we were dividing a whole number by a fraction. However, the same method applies equally to dividing a fraction by a fraction.

Example

What is $\frac{1}{2} \div \frac{1}{4}$?

$$\frac{1}{2} \div \frac{1}{4} \quad = \quad \frac{1}{2} \times \frac{4}{1} \quad = \quad \frac{4}{2} \quad = \quad 2$$

Dividing mixed fractions

To divide mixed fractions, change them to top-heavy fractions and then use the above method.

Practice 14

1. $\frac{1}{4} \div \frac{1}{8}$

2. $\frac{1}{3} \div \frac{2}{9}$

3. $\frac{3}{5} \div \frac{1}{10}$

4. $1\frac{1}{2} \div 2\frac{3}{4}$

5. $6\frac{3}{4} \div 3\frac{1}{2}$

Fraction problems

Fraction calculations often seem more difficult if they are presented as written questions. The words often obscure the calculation that has to be done.

The advice given for tackling whole number problems also applies to problems involving fractions:

 Read the question carefully.

 Identify those parts of the question that relate to the calculation you need to carry out.

Use a highlighter pen if it helps.

 Look for key words such as fraction **of** a number, one number **as a fraction of** another number, find the **fraction left**, what is the **total fraction** … to help you identify what type of fraction calculation you are required to do.

 Write out the important information more simply, using appropriate numerical symbols if possible.

Example

John earns £400 a week. He spends a fifth $(\frac{1}{5})$ on rent, a quarter $(\frac{1}{4})$ on food and a tenth $(\frac{1}{10})$ on bills.

What is the total fraction spent on these items?

How much does he spend on these items?

What fraction is left for all his other requirements?

Highlighting important information:

John **earns £400 a week**. He spends $\frac{1}{5}$ on rent, $\frac{1}{4}$ on food and $\frac{1}{10}$ on bills.

What is the **total fraction** spent on these items?

How much does he spend on these items?

What fraction is left for all his other requirements?

Writing out the information more simply:

Total earnings: £400

$\frac{1}{5}$ on rent

$\frac{1}{4}$ on food

$\frac{1}{10}$ on bills

Identify key words – **total fraction, how much does he spend, what fraction is left** – and write down the type of calculation you think this involves.

total fraction – add up the fractions
how much does he spend? – find that fraction of his total money
what fraction is left? – subtract that fraction from a whole (1)

Finally, carry out the calculations:

total fraction

$$\frac{1}{5} + \frac{1}{4} + \frac{1}{10} \quad = \quad \frac{4}{20} + \frac{5}{20} + \frac{2}{20} \quad = \quad \frac{11}{20}$$

So John spends $\frac{11}{20}$ of his income on these three items.

how much does he spend?

$$\frac{11}{20} \text{ of } £400 = \frac{11}{20} \times £400 = \frac{11}{1\,20} \times £400\ 20 = \frac{£220}{1} = £220$$

He spends £220 on these three items.

what fraction is left?

$$1 - \frac{11}{20} \quad = \quad \frac{20}{20} - \frac{11}{20} \quad = \quad \frac{9}{20}$$

He has $\frac{9}{20}$ of his money left.

Practice 15

1. Every month, Luke gives $\frac{1}{10}$ of his income to charity. Last month he earned £1440. How much did he give to charity?

2. Out of an adult numeracy class of 20 students, 12 were women. What fraction were women? Write your answer in its simplest form.

3. A doctor has 186 patients over sixty years of age. She has an annual target of giving the 'flu' vaccination to $\frac{2}{3}$ of these. Last year she vaccinated 105 patients. Did she meet her target?

4. In a survey of students, $\frac{1}{4}$ said they regularly drank more than the recommended number of units of alcohol each week. $\frac{2}{3}$ said they drank regularly but never more than the recommended number of units. The remaining students said they never drank alcohol. What fraction said they never drank alcohol?

5. A kilogram of cheese costs £5. What would be the cost of $2\frac{1}{2}$ kilograms of cheese?

Fractions on the calculator

Most numeracy calculators can't display fractions, so carrying out fraction calculations is impossible. However, most scientific calculators have a function that enables you to enter a fraction. If you can enter a fraction into your calculator then all fraction calculations can be done with the calculator.

The fraction button on a scientific calculator looks like this:

To use this function, enter the top number of the fraction, press  and then enter the bottom number of the fraction.

For example, if you enter $\frac{2}{3}$ then the display will show:

$$2 \lrcorner 3$$

The $\lrcorner$ symbol is how most scientific calculators show one number over another, but some modern scientific calculators are able to show fractions in the way they are written:

$$\frac{2}{3}$$

Mixed fractions can be entered in a similar way. For example, to enter $1\frac{1}{2}$ you would press:

The display shows:

$$1 \lrcorner 1 \lrcorner 2$$

If you are able to enter a fraction into your calculator then you can then use the × ÷ + − functions to carry out calculations. However, it is still a good idea to make sure that you understand what is happening by practising doing the calculations by hand as well.

DECIMALS

What are decimals?

The decimal system gives us an alternative to using fractions when we describe parts of a whole number.

 Decimals are also sometimes called decimal fractions.

The history of the development of decimals goes back centuries. Some historians think that the Arabic mathematician al-Uglidisi (AD 952) invented them. Others believe that the idea came originally from the Indian subcontinent.

In the section about whole numbers, we saw that our number system is based on counting in groups of ten. For example, the number 123 represents 1 hundred plus 2 tens plus 3 ones.

$$123 = 100$$
$$+ 20$$
$$+ 3$$

Each position in the number, moving from right to left, has a place value ten times bigger than the one before it.

The decimal fraction system extends this idea so that it includes numbers less than 1. This is how it is structured:

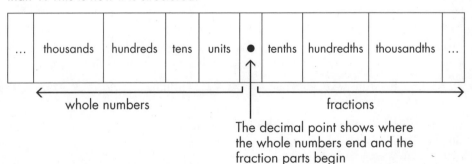

| ... | thousands | hundreds | tens | units | • | tenths | hundredths | thousandths | ... |

whole numbers ← → fractions

The decimal point shows where the whole numbers end and the fraction parts begin

Decimals

Writing decimals

We can use the diagram of the decimal fraction system to see how to write tenths, hundredths, thousandths, etc, as decimals:

The number $1\frac{5}{10}$ would be written as 1.5 – one unit and five tenths.

The number $12\frac{4}{100}$ would be written as 12.04 – one ten, two units, no tenths and four hundredths.

The number $\frac{7}{10}$ would be written as 0.7 – no units and seven tenths.

We can also change decimals into numbers containing tenths, hundredths, thousandths, etc:

The decimal number 3.2 means 3 units and 2 tenths, or $3\frac{2}{10}$.

The decimal number 50.06 means five tens, no units, no tenths and six hundredths, or $50\frac{6}{100}$.

The decimal number 123.007 means one hundred, two tens, three units, no tenths, no hundredths and seven thousandths, or $123\frac{7}{1000}$.

We will see later in this section that calculations with decimals can be done without having to know what each decimal number means as a fraction. However, an understanding of how the decimal system works will help you to make sense of what you are doing.

Two further examples will illustrate a point that sometimes causes confusion:

Example

Write $2\frac{17}{100}$ as a decimal.

The seventeen hundredths cannot be 'fitted into' the column for hundredths, as this column can take only one figure. It has to 'spill over' into the tenths column.

This is fine, because seventeen hundredths – $\frac{17}{100}$ – is the same as $\frac{1}{10}+\frac{7}{100}$ – one tenth and seven hundredths.

So, the mixed fraction $2\frac{17}{100}$ – two units and seventeen hundredths – as a decimal number is written:

2.17

Example

Write $\frac{125}{1000}$ as a decimal.

The one hundred and twenty-five thousandths cannot be 'fitted into' the column for thousandths, as this column can take only one figure. It has to 'spill over' into the hundredths and tenths columns.

This is fine, because one hundred and twenty-five thousandths – $\frac{125}{1000}$ – is the same as $\frac{1}{10} + \frac{2}{100} + \frac{5}{1000}$ – one tenth, two hundredths, and five thousandths.

So, the fraction $\frac{125}{1000}$ – one hundred and twenty-five thousandths – is written:

0.125

In practice, there is no need to break a fraction down like this when writing it as a decimal. In the later section on the connection between fractions, decimals and percentages we will look at a quick method of writing fractions as decimals and vice versa.

Decimals and other fractions

It might seem from the explanation of how the decimal fraction system works that it can only be used when working with tenths, hundredths, thousandths, and so on. However, this is not the case. The decimal fraction system can be used to represent *any* fraction. This is because any fraction can be broken down into tenths, hundredths, thousandths and so on. The next two examples illustrate this:

Example

Write $\frac{1}{2}$ as a decimal.

$\frac{1}{2} = \frac{5}{10}$ (see the section on page 43 about equivalent fractions)

$\frac{5}{10} = 0.5$

So $\frac{1}{2}$ can be written as 0.5.

Example

Write $\frac{3}{4}$ as a decimal.

The fraction $\frac{3}{4}$ can in fact be broken down into seven tenths and five hundredths:

$$\frac{3}{4} = \frac{7}{10} + \frac{5}{100}$$

(You don't need to know how this works, but you can check that it is true by adding the two fractions together – use the method shown on page 57 in the section about adding fractions.)

$$\frac{7}{10} + \frac{5}{100} = 0.75$$

So $\frac{3}{4}$ can be written as 0.75.

Again, in practice, there is no need to break a fraction down like this when writing it as a decimal. In the later section on the connection between fractions, decimals and percentages we will look at a much quicker and simpler method.

Using zeros in decimal numbers

Notice that, in the last example, the decimal number 0.75 is written with a zero before the point. This zero is written not so much to indicate that there are no units in the number, but to draw attention to the decimal point. Without the zero it would appear as .75 – the decimal point could easily be overlooked and the number taken to be 75 (seventy-five).

 It is good practice always to write a zero before a decimal number that has no whole numbers.

The decimal number one and five tenths could be written as 1.5 or 1.50 or indeed 1.500000. The zeros at the end of a decimal number are optional, and are usually left out.

 It is unnecessary to write zeros at the end of a decimal fraction.

Adding decimals

Decimal numbers are added in much the same way as whole numbers. With whole numbers it was important to make sure that the numbers to be added were written one under the other so that units lined up with units, tens with tens, hundreds with hundreds and so on. When adding decimal numbers we also have to make sure that tenths line up with tenths, hundredths with hundredths, thousandths with thousandths and so on.

Example

Jacob wanted to fence off three sides of his garden. He measured the sides and found that they were 15.35 metres, 7.20 metres and 10.75 metres long respectively. What total length of fencing did Jacob need?

Adding the three lengths together:

$$
\begin{array}{r}
15.35 \\
7.20 \\
+\ \ \underline{10.75} \\
33.30
\end{array}
$$

The numbers are written so that the tens, the units, the tenths and the hundredths line up

Jacob needed 33.3 metres of fencing. (Notice how the last zero can be left out.)

In practice, you don't have to worry about lining up the hundreds, tens, ones, tenths, hundredths, and so on. As long as the decimal points are lined up under each other then the other numbers will automatically be in the correct positions.

Adding a whole number to a decimal

It can be a little confusing when adding a whole number to a decimal because usually the whole number does not have a decimal point.

Example

What is 25 + 3.45?

Any whole number can be written with a decimal point immediately to the right of the units. When doing this it is good practice to put a zero after the decimal point (in the tenths position) to emphasize the decimal point.

So we can write 25 as 25.0

Now, when adding, the decimal points can be lined up as above.

$$
\begin{array}{r}
25.0 \\
+\ \ \underline{3.45} \\
28.45
\end{array}
$$

Decimals

Carrying

Where necessary, we can carry from one position to the next in the same way as when adding whole numbers (see page 8 in the section about whole numbers).

1.	45.68 + 1.38
2.	0.28 + 92.931
3.	1.005 + 9.789
4.	0.006 + 18.901
5.	39.0012 + 567.21 + 0.0213

Subtracting decimals

When subtracting decimals, the decimal points in each number are lined up under each other as described above.

Example

Scott weighs 68.75 kilograms. Mairi weighs 56.25 kilograms. How much heavier is Scott than Mairi?

We need to find the difference between their weights.

$$\begin{array}{r} 68.75 \\ -\ \underline{56.25} \\ 12.50 \end{array}$$

Scott is 12.5 kilograms heavier.

Borrowing and carrying

We can borrow and carry in the same way as for subtracting whole numbers (see page 15 in the section about whole numbers).

Practice 2

1.   8.24 – 6.12
2.   65.356 – 13.12
3.   9.35 – 5.19
4.   0.536 – 0.077
5.   12.501 – 3.294

Multiplying decimals

Multiplying a decimal by a whole number

Multiplying a decimal by a whole number is done in almost the same way as multiplying two whole numbers.

Example

Krishna wants to buy a length of wood to make six shelves. Each shelf is 1.5 metres long. How much wood will she need in total?

We need to calculate 1.5 × 6. The calculation can be laid out as follows:

$$
\begin{array}{r}
1.5 \\
\times \quad \underline{6}
\end{array}
$$

Notice that the 6 we are multiplying by has been placed under the first figure on the right hand side. In effect, we are treating the decimal number as a whole number. It is fine to do this, as long as we put the decimal point in the answer directly under the decimal point in the calculation.

$$
\begin{array}{r}
1.5 \\
\times \quad \underline{.6} \\
9.0
\end{array}
$$

The dotted line shows how the decimal point in the answer lines up with the decimal point in the calculation. (The carrying isn't shown; if you're unsure then see page 8 in the section about whole numbers.)

Decimals

The answer is 9.0 or 9, so Krishna will need 9 metres of wood for her shelves.

Multiplying by larger whole numbers

The same principle applies when multiplying by larger whole numbers.

Example

Calculate 2.75 × 15.

The calculation can be laid out like this:

$$
\begin{array}{r}
2.75 \\
\times \quad 15 \\
\hline
1375 \\
275 \\
\hline
41.25
\end{array}
$$

The answer is 41.25 – the dotted line shows how the decimal point has been placed in the answer under the point in the calculation. (Again, the carrying has been omitted.)

Practice 3

1. 2.8 × 5
2. 6.23 × 7
3. 0.45 × 12
4. 23.02 × 25
5. 18 × 0.25 (Hint: turn the multiplication around and multiply 0.25 by 18)

Multiplying by 10, 100, 1000 ...

If we multiply the number 1.25 by 10 using the method described above, the answer is 12.5 – it appears that the decimal point has moved one place to the right.

If we multiply the number 2.125 by 100 using the same method, the answer is 212.5 – it appears that the decimal point has moved two places to the right.

If we multiply the number 0.12753 by 1000 using the same method, the answer is 127.53 – it appears that the decimal point has moved three places to the right.

This gives us a quick method for multiplying decimal numbers by 10, 100, 1000 and so on.

 To multiply a decimal number by 10, 100, 1000 and so on, move the decimal point one place to the right for every zero in the number you are multiplying by. Move the decimal point once to the right when multiplying by 10, twice when multiplying by 100, three times when multiplying by 1000 and so on.

Multiplying a decimal by a decimal

The technique of keeping the decimal points in a straight line doesn't work when multiplying two decimal numbers.

Example

Calculate 0.5 × 2.0.

$0.5 = \frac{5}{10} = \frac{1}{2}$, and 2.0 is the same as 2, so this calculation is the same as $\frac{1}{2} \times 2$.

$\frac{1}{2} \times 2 = 1$ (a half of 2 is 1)

So the answer to the calculation 0.5 × 2.0 is 1.

Have a look at what happens if we try to do the calculation by keeping the decimal points in a straight line:

$$
\begin{array}{r}
2.0 \\
\times \quad \underline{0.5} \quad \textbf{✗} \\
10.0
\end{array}
$$

This is obviously wrong – we know that the answer should be 1. The answer is wrong because the decimal point is in the wrong place.

To avoid confusion with decimal points when multiplying two decimal numbers it is better to treat both decimal numbers as whole numbers (in effect, ignoring the points altogether). Carry out the calculation using these whole numbers, and put the decimal point back into the answer in the correct place afterwards.

Example

Calculate 23.42 × 5.6.

Carry out the calculation without the decimal points:

$$
\begin{array}{r}
2342 \\
\times \quad 56 \\
\hline
14052 \\
117100 \\
\hline
131152
\end{array}
$$

(The carrying has not been shown.)

Now use the following method to decide where to put the decimal point:

 To decide where to put the decimal point, count the total number of **decimal figures** in the original calculation.

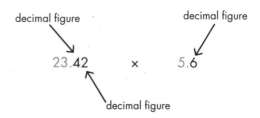

There are 3 decimal figures in this calculation, so there should be three decimal figures in the answer. Counting from the right-hand side:

131.152

The answer is 131.152.

> ## Practice 4
>
> 1. 2.5×1.2
> 2. 1.23×1.2
> 3. 4.516×0.8
> 4. 20.5×0.15
> 5. 0.003×0.00006

Dividing decimals

Dividing a decimal by a whole number

Dividing a decimal by a whole number is done in almost the same way as dividing two whole numbers.

> ## Example
>
> A piece of copper pipe 2.4 metres long is to be cut into four equal pieces. What will be the length of each piece?

We need to calculate $2.4 \div 4$:

$$\begin{array}{r} 0.6 \\ 4\overline{\smash{)}2.4} \end{array}$$

(The carrying has not been shown – if you need a reminder, see page 28 in the section about whole numbers.)

The decimal point in the answer is placed directly above the decimal point in the number in the division box.

Each piece of copper pipe would be 0.6 metres long.

Dealing with remainders

The next example shows what to do with remainders when dividing into a decimal.

Decimals

Example

Kenneth's allotment is 31.5 metres long. He wants to divide it into four equal sections. What will be the length of each section?

We need to divide 31.5 by 4:

$$4 \overline{\smash{\big)}\ 3\ 1\ .\ 5}$$

4 into 3 won't go, so we move to the next figure and see how many times 4 will go into 31.

4 into 31 goes seven times (7 × 4 = 28) with 3 left over, so we must carry the 3 to the next figure:

$$4 \overline{\smash{\big)}\ 3\ 1\ .^{3}5} \quad \begin{array}{c} 7 \\ \end{array}$$

4 goes into 35 eight times (8 × 4 = 32) with 3 left over, so we must carry the 3 to the next figure:

$$4 \overline{\smash{\big)}\ 3\ 1\ .^{3}5} \quad \begin{array}{c} 7\ .\ 8 \\ \end{array} \quad \rightarrow 3?$$

It seems that there is no figure to which we can carry this last remainder, but with decimal numbers we can add as many zeros as we want to the end of the number and then continue dividing.

Adding one zero and carrying the 3:

$$4 \overline{\smash{\big)}\ 3\ 1\ .^{3}5^{3}0} \quad \begin{array}{c} 7\ .\ 8 \\ \end{array}$$

4 goes into 30 seven times (7 × 4 = 28) with 2 left over, so we must carry the 2 to the next figure:

$$4 \overline{\smash{\big)}\ 3\ 1\ .^{3}5^{3}0} \quad \begin{array}{c} 7\ .\ 8\ 7 \\ \end{array} \quad \rightarrow 2?$$

Again, it seems that there is no figure to which we can carry the 2, so we add another zero:

$$7.87$$
$$4 \overline{) 31.5\overset{3}{5}\overset{3}{0}\overset{2}{0}}$$

4 goes into 20 five times exactly. There is no remainder, so this is the end of the calculation.

$$7.875$$
$$4 \overline{) 31.5\overset{3}{5}\overset{3}{0}\overset{2}{0}}$$

So 7.875 is the answer.

Sometimes, as in this example, adding one or more zeros will allow you to complete the division. Sometimes, no matter how many zeros are added, there is still a remainder to be carried. When that happens, you have to decide when you should stop dividing. This will depend on how many decimal places you want in your answer; see the section below about rounding off.

Practice 5

1. $8.32 \div 4$
2. $12.25 \div 5$
3. $1.44 \div 12$
4. $0.0056 \div 7$
5. $234.63 \div 8$

Dividing by 10, 100, 1000 ...

If we divide the number 12.5 by 10 using the method described above, the answer is 1.25. It appears that the decimal point has moved one place to the left.

If we divide the number 345.5 by 100 using the method described above, the answer is 3.455. It appears that the decimal point has moved two places to the left.

If we divide the number 5002.5 by 1000 using the method described above, the answer is 5.0025. It appears that the decimal point has moved three places to the left.

This gives us a quick method for dividing decimal numbers by 10, 100, 1000 and so on:

 To divide a decimal number by 10, 100, 1000 and so on, move the decimal point one place to the left for every zero in the number you are dividing by. Move the decimal point once to the left when dividing by 10, twice when dividing by 100, three times when dividing by 1000 and so on.

Dividing a decimal by a decimal

First think about the whole number division 8 ÷ 2.

In this calculation we are trying to work out how many twos are in eight. The answer is 4.

$$8 \div 2 = 4$$

Now think about the decimal division 1.5 ÷ 0.5.

$$1.5 \div 0.5 \quad = \quad 1\tfrac{1}{2} \div \tfrac{1}{2}$$

In this calculation, we are trying to work out how many halves are in one and a half. This is a bit more difficult, but knowing that there are two halves in one whole, we can work out that there must be three halves in one and a half.

$$1.5 \div 0.5 \quad = \quad 3$$

Now think about the division 1.5 ÷ 0.01.

Here we trying to work out how many hundredths are in one and a half. This is even more difficult. In general, except in a few cases, division of a decimal by another decimal is not easy to do.

Instead, we use the fact that it is possible to move the decimal points in a calculation and get the same answer.

For example:

$$1.25 \div 0.05$$

... is the same as:

12.5 ÷ 0.5 (both decimal points moved one place to the right)

… which is the same as:

125 ÷ 5   (both decimal points moved another place to the right)

The fact that we are moving the decimal point *the same number of places* in each number keeps the calculation the same.

We can use this idea to simplify decimal division:

 To divide a decimal number by another decimal number, make the number you are dividing by into a whole number by moving the decimal point. Then move the decimal point in the other number the same number of times.

Example

Calculate 14.21 ÷ 0.7.

First move the decimal point one place to the right in the divider:

0 . 7   becomes 07, which is the same as 7

Then move the decimal point one place in the other number:

1 4 . 2 1   becomes 142.1

The division becomes 142.1 ÷ 7:

$$\begin{array}{r} 20.3 \\ 7\,\overline{)\,142.1} \end{array}$$

Decimals

1. $25.25 \div 0.5$
2. $918.9 \div 0.9$
3. $0.144 \div 1.2$
4. $12.2504 \div 0.008$
5. $0.0056 \div 0.0007$

The next example shows what to do if the number you are dividing by has more decimal places than the number you are dividing into.

Example

$0.125 \div 0.0005$.

To make 0.0005 a whole number, we have to move the decimal point four places to the right. This turns 0.0005 into 5.

0 . 0 0 0 5 becomes 5

Next, we try to move the decimal point in 0.125 by four places as well. However, we need to add a zero to the end of the number so that we have enough decimal places to move the decimal point.

0 . 1 2 5 = 0 . 1 2 5 0 which becomes 01250 or 1250

In this example, we had to add one zero to make it possible to move the decimal point the same number of times as we did in the number we were dividing by. In practice, we can add as many zeros as we need to.

Using a calculator

If you find handling decimals difficult, you can use a calculator to work with them.

All the above operations can be carried out on a calculator using the +, −, × and ÷ function buttons. The decimal point can be put into numbers using the ⬚ function button.

 Care should be taken when entering decimal numbers into a calculator. It's easy to misplace the decimal point, making the calculation incorrect. It's a good idea to double-check each number after it has been entered.

Rounding off

When carrying out decimal calculations we sometimes get answers with more decimal figures than we need.

Example

A piece of material 5 metres long is to be cut into seven equal pieces. How long will each piece be?

Dividing 5 by 7 either on a calculator or on paper gives an answer of:

$5 \div 7 = 0.714285714 \ldots$

The actual number of decimal figures in the answer will depend on the size of the display on the calculator, or on your patience if doing the calculation by hand.

It doesn't make sense to use all these decimal figures. The first two decimal places in the answer represent centimetres. The third decimal place represents millimetres (see page 150 in the section about measurement). Using a tape measure or a ruler, it would be impossible to measure more accurately than this, so there would be no point in using any more than three decimal figures.

So a common sense answer to the division would be 0.714 metres.

The process of shortening a decimal answer is sometimes referred to as rounding off. There are two ways of rounding off a decimal number.

Rounding off to a given number of decimal places or figures

 For one decimal place, use just the first figure after the decimal point.

 For two decimal places, use just the first two figures after the decimal point.

 For three decimal places, use just the first three figures after the decimal point.

 For four decimal places, use just the first four figures after the decimal point.

However, when rounding to 1 decimal place, always check the second decimal place. If this is 5 or more then round up the first decimal figure by 1.

5.**35** to 1 decimal place is 5.**4**

When rounding to 2 decimal places, always check the third decimal place. If this is 5 or more then round up the second decimal figure by 1.

12.3**65** to 2 decimal places is 12.3**7**

When rounding to 3 decimal places, always check the fourth decimal place. If this is 5 or more then round up the third decimal figure by 1.

7.42**37**9 to 3 decimal places is 7.42**4**

If rounding to 4 decimal places, always check the fifth decimal place. If this is 5 or more then you round up the fourth decimal figure by 1.

0.351**28** to 4 decimal places is 0.351**3**

When rounding to a given number of decimal places, always check the next decimal figure to the right. If it is 5 or more, the last decimal place that you are using must be rounded up.

Practice 7

Round off the following decimal numbers to the required number of decimal places:

1. 0.324 to 2 decimal places
2. 17.3456 to 3 decimal places
3. 4.8292 to 2 decimal places
4. 12.09057 to 1 decimal places
5. 0.20935 to 3 decimal places
6. 8.0737 to 2 decimal places
7. 18. 0036 to 3 decimal places
8. 0.005 to 2 decimal places
9. 250.995 to 1 decimal place
10. 3.142857143 to 5 decimal places

Significant figures

In any number, some figures can be considered more **significant** or important than others.

For example, imagine that you have won £3 167 891 on the National Lottery. When a friend asks you how much you have won, you might say 'Over three million pounds'. The three million is more **significant** or important than the other figures.

In general, the bigger the figure, the more significant it is.

 In some books you will see 'significant figures' abbreviated to 'sig. figs'.

Here's how to round off to a given number of significant figures:

 For 1 significant figure, use the first figure on the left of the number and put zeros in place of the other numbers.

For 2 significant figures, use the first two figures on the left of the number and put zeros in place of the other numbers.

 For 3 significant figures, use the first three figures on the left of the number and put zeros in place of the other numbers.

 For 4 significant figures, use the first four figures on the left of the number and put zeros in place of the other numbers.

However, when rounding to 1 significant figure, always check the second figure. If this is 5 or more then round up the first significant figure by 1.

85765 to 1 significant figure is **90**000

When rounding to 2 significant figures, always check the third figure. If this is 5 or more then round up the second significant figure by 1.

85**7**65 to 2 significant figures is 86**000**

When rounding to 3 significant figures, always check the fourth figure. If this is 5 or more then round up the third significant figure by 1.

85**765** to 3 significant figures is 85**8**00

When rounding to 4 significant figures, always check the fifth figure. If this is 5 or more then round up the fourth significant figure by 1.

85**765** to 4 significant figures is 85**7**70

 When rounding to a given number of significant figures, always check the next figure to the right. If it is 5 or more, the last significant figure that you are using must be rounded up.

And there is one more thing to check:

It is important to remember to put zeros in place of any figures not being used. Leaving them out will make the answer incorrect.

For example, returning to the imaginary lottery win of £3 167 891:

To 1 significant figure this is written as £3 000 000 (three million pounds). However, if we forget to put in the zeros, the amount of money to 1 significant figure would be £3 (three pounds), obviously not a good approximation of the lottery win!

Significant figures and decimal zeros

In general, when writing decimal numbers to a given number of significant figures, you should follow the rules described above. However, care must be taken with any zeros in the decimal part of a number, as some may be significant and some not.

 Zeros **'inside'** the decimal part of a number – those that are neither at the beginning nor at the end of the decimal part – are **always significant**.

 Zeros at the **beginning or end** of the decimal part of a number, on the other hand, **should not** be counted as significant.

 If the number has no whole number part, then the zero before the decimal point is not significant.

A few examples will clarify this:

Example

What is 0.506 to 2 significant figures?

The zero 'inside' the decimal part of the number is significant and should be counted when rounding off. It should be rounded up because the next figure – the 6 – is greater than or equal to 5. The zero before the point is not significant.

0.506 to 2 significant figures is 0.51

Decimals

Example

What is 0.0632 to 1 significant figure?

The zero before the point and the zero at the beginning of the decimal part of the number are not significant and should not be counted when rounding off. This means that the 6 is the first significant figure. The next figure, the 3, is less than 5 so we do not need to round up the 6.

0.0632 to 1 significant figure is 0.06

Example

What is 0.0109 to 2 significant figures?

The zero at the beginning of the decimal part is not significant. However, the zero 'inside' the decimal is significant, and should be rounded up because the next figure – the 9 – is greater than or equal to 5.

0.0109 to 2 significant figures is 0.011

Practice 8

Round off the following numbers to the required number of significant figures:

1. 345 to 2 significant figures

2. 1554 to 3 significant figures

3. 83 to 1 significant figure

4. 12.567 to 2 significant figures

5. 0.628682 to 5 significant figures

6. 0.3056 to 3 significant figures

7. 0.00057 to 1 significant figure

8. 60.40078 to 5 significant figures

9. 0.0097 to 1 significant figure

10. 733.01 to 2 significant figures

MONEY

The British system of money

The British system of money was **decimalized** on 15th February 1971. This means that British money can be added, subtracted, divided and multiplied in the same way as ordinary decimal numbers.

There are two units of currency used in the British system of money: **pounds** (£) and **pence** or **pennies** (p).

Pounds are written with a **£** sign in front of the number. For example, five pounds is written as £5.

Pence are written with a small **p** after the number. For example, twenty-five pence is written as 25p.

 There are a hundred pence in a pound: 100p = £1

Because there are 100 pence in one pound, 25 pence can also be thought of as $\frac{25}{100}$ of a pound. This means it can also be written as the decimal £0.25 (see the section on page 76 about changing fractions to decimals). Any number of pence can be written as a decimal part of a pound:

54p = £0.54
35p = £0.35
10p = £0.10

 Notice that when pence are written as a decimal, we have to put a £ sign at the front to show that we have changed the pence into a decimal part of a pound.

If the number of pence is lower than 10p, then the decimal equivalent will have a zero after the decimal point. For example, 5p would be written as £0.05 (5p is $\frac{5}{100}$ of a pound).

1p	=	£0.01
2p	=	£0.02
3p	=	£0.03
4p	=	£0.04

Money

5p	=	£0.05
6p	=	£0.06
7p	=	£0.07
8p	=	£0.08
9p	=	£0.09
10p	=	£0.10

An amount of money containing both pounds and pence is written as a decimal number with a pound sign (£). For example, eight pounds and seventy-four pence is written as £8.74. Notice that if we are writing an amount containing pounds and pence, we do not put the pence sign at the end.

Operations with money

The operations of adding, subtracting, multiplying and dividing amounts of money are carried out in the same way as shown in the decimal section of this book. Read through the following examples and try the practice exercises. You can refer back to the section on decimals to refresh your memory if necessary.

Example

At the supermarket Fatima spent £2.60 on apples, £1.90 on oranges and £0.98 on bananas. How much did she spend in total?

To find the total we add the three amounts:

$$
\begin{array}{r}
£2.60 \\
£1.90 \\
+ \; £0.98 \\
\hline
£5.48 \\
\end{array}
$$

So Fatima spent £5.48 in total.

Example

At the cinema Edwin pays for his ticket with a £10 note. If the ticket costs £4.75, how much change does he get?

To calculate Edwin's change we need to subtract £4.75 from £10. Notice how the £10 is written with two zeros after the decimal point to make subtraction possible.

$$
\begin{array}{r}
£\ \overset{9}{\cancel{1}}\overset{9}{\cancel{0}}\cancel{0}0 \\
-\ £\ \ \ 4.75 \\
\hline
£\ \ \ 5.25
\end{array}
$$

Edwin gets £5.25 change.

Example

Sean pays his gas bill by monthly direct debit. Each payment is £20.65. How much does he pay each year?

There are twelve months in a year, so we multiply £20.65 by 12.

$$
\begin{array}{r}
£\ \ \ 20.65 \\
\times\ \ \ \ \ \ \ 12 \\
\hline
£\ \ \ 41.30 \\
£\ 206.50 \\
\hline
£\ 247.80
\end{array}
$$

Sean pays £247.80 each year.

Example

£273 is shared between 4 people. How much does each person get?

To work out each person's share, we need to divide the total amount by 4. Notice how the £273 is written with two zeros after the decimal point to make the division possible.

$$
\begin{array}{r}
£\ \ 68.25 \\
4\ \overline{)\,£\,27^{3}3.\,{}^{1}0\,{}^{2}0}
\end{array}
$$

Each person's share is £68.25.

Practice 1

1.	£31.45 + £76.90	6.	£719.08 – £27.56
2.	£345.32 + £7.86 + £123.80	7.	£153.82 × 6
3.	£39.50 + 56p	8.	£98.32 × 12
4.	£92.29 – 74p	9.	£982.20 ÷ 5
5.	£58.45 – £23.22	10.	£83.16 ÷ 12

Financial numeracy

Calculations with money are needed in many everyday situations. It is beyond the scope of this book to look at all of them, so we will look at just two examples.

Currency conversion

The type of money used in each country is called the **currency**. It can be useful to change British pounds and pence into another currency. To do this, we need to know the **exchange rate**.

 The exchange rate tells us how much of a foreign currency we will get for £1. It can vary from day to day and also depends on where you change your money.

The following examples illustrate how to convert from one currency to another:

Example

Richard is going on holiday to Spain. He changes £500 into euros, the currency of Spain. How many euros does he get if the exchange rate is £1 = 1.6 euros (1.6 euros to the pound)?

To work this out, we need to multiply the exchange rate by the amount of money that is being changed.

£1	=	1.6 euros		
£500	=	1.6 euros × 500	=	800 euros

Richard gets 800 euros.

To change British currency into a foreign currency, multiply the amount of British money by the exchange rate.

Example

Louise has been on holiday in America. She returns to Britain with $200 (200 dollars), which she wants to change back into pounds. How many pounds will she get if the exchange rate is £1 = $1. 75?

Louise will get £1 for every $1.75 that she has, so we need to calculate how many '1.75 dollars' are in the $200. To do this we divide the $200 by 1.75

$200 ÷ 1.75 = £114.29 to the nearest penny

Louise will get £114.29.

 To change a foreign currency into British currency, divide the amount of foreign currency by the exchange rate.

Practice 2

Change the British currency in the questions below into the currency required, using the exchange rate quoted. Give your answers to the nearest penny.

1. Change £300 into euros at an exchange rate of 1.5 euros to the pound.

2. Change £850 into euros at an exchange rate of 1.6 euros to the pound.

3. Change £1200 into US dollars at an exchange rate of 1.8 dollars to the pound.

4. Change £750 into Danish kroner at an exchange rate of 10.5 kroner to the pound.

5. Change £620 into Indian rupees at an exchange rate of 78 rupees to the pound.

Practice 3

Change the foreign currency in the questions below into British currency using the exchange rate quoted. Give your answers to the nearest penny.

1. Change 3000 South African rand into pounds at an exchange rate of 12.2 rand to the pound.

2. Change 950 US dollars into pounds at an exchange rate of 1.8 dollars to the pound.

3. Change 1500 Maltese liri into pounds at an exchange rate of 0.6 liri to the pound.

4. Change 220 euros into pounds at an exchange rate of 1.4 euros to the pound.

5. Change 50 Swiss francs into pounds at an exchange rate of 2.2 francs to the pound.

Exchange rates are often long numbers with several decimal places. If you need to convert currencies in real life, you may find it easier to use a calculator to do the multiplying or dividing.

Savings

If you save money in a savings account, then you are usually paid **interest**.

Interest is the money the bank adds to your savings for the privilege of using your money while it is in the savings account.

Equally, if you borrow money from a bank or mortgage lender then you are charged interest on the loan.

The bank or mortgage company charges you interest for the privilege of borrowing money from them.

The amount of interest paid will depend on the **interest rate** being paid or charged by the bank at the time. The interest rate is usually expressed as a percentage per annum (each year). Sometimes, however, the interest is paid monthly, and then the interest rate is expressed as a monthly percentage.

The following examples show how the interest is calculated for different periods of time.

Example

Rashid puts £500 into a savings account paying 3% per annum. If he leaves it in for a year, how much interest does he earn?

Interest paid in one year = 3% of £500 = $\frac{3}{100}$ of £500 = (£500 ÷ 100) × 3
= £5 × 3 = £15

Rashid receives £15 interest in one year.

Example

Alison puts £2000 into a savings account paying 5% per annum. She withdraws her money after six months. How much interest does she earn?

First find the interest paid in one year:

Interest paid in one year = 5% of £2000 = $\frac{5}{100}$ of £2000

= (£2000 ÷ 100) × 5 = £20 × 5 = £100

The interest paid in six months will be half of this:

Interest paid in six months = £100 ÷ 2 = £50

Alison earns £50 in interest.

 Banks sometimes use a different method to calculate the interest for part of a year, which gives a slightly lower amount. However, it is more complicated and the difference is very small, so it is easier to use the method described above.

The final example shows how banks calculate interest on money saved for a number of years. The important thing to notice in this example is how the interest from each year is added on to the amount in the savings account before the interest is calculated for the following year. This process is often referred to as compound interest.

Example

Sandy puts £1200 into a savings account paying 6% per annum. She withdraws her money after 3 years. How much interest does she earn?

Year 1:

Interest earned = 6% of £1200 = $\frac{6}{100}$ × £1200 = (£1200 ÷ 100) × 6
= £12 × 6 = **£72**

Year 2:

Before calculating the interest for year 2, the interest earned in year 1 must be added on.

£1200 + £72 = £1272

Now the interest for year 2 can be found:

Interest earned = 6% of £1272 = $\frac{6}{100}$ × £1272 = (£1272 ÷ 100) × 6
= £12.72 × 6 = **£76.32**

Year 3:

Before calculating the interest for year 3, the interest earned in year 2 must be added on.

£1272 + £76.32 = £1348.32

Now the interest for year 3 can be found:

Interest earned = 6% of £1348.32 = $\frac{6}{100}$ × £1348.32
= (£1348.32 ÷ 100) × 6 = £13.48 × 6 = **£80.88**

The total interest over the three years will be the sum of the interests from year 1, year 2 and year 3:

Total interest earned = £72 + £76.32 + £80.88 = **£229.20**

Sandy earns £229.20 in interest.

When tackling problems about interest, read the questions carefully. You may be asked to calculate the interest earned, as in the above examples, or you may be asked to calculate the final amount of money in a savings account. In that case, the total interest earned has to be added on to the original amount deposited in the account.

Practice 4

Find the interest earned on the following savings, giving your answers to the nearest penny:

1. £400 at 2% for 1 year
2. £2000 at 4% for 2 years
3. £850 at 5% for 3 years
4. £1900 at 6% for 6 months
5. £300 at 7% for 3 months

Practice 5

Find the final amount in the following savings accounts, giving your answers to the nearest penny:

1. £600 for 3 years at 3%
2. £345 for 2 years at 5%
3. £780 for 4 years at 2%
4. £1600 for 3 years at 7%
5. £900 for 6 months at 3%

PERCENTAGES

What are percentages?

A percentage, like a fraction or a decimal, is a way of describing part of a whole.

You might recognize the percentage 50% (fifty per cent) as being the same as the fraction a half ($\frac{1}{2}$), or the decimal 0.5 (nought point five).

Percentages tend to be a more 'user-friendly' way of describing parts of a whole. They are based on the idea that a whole can be divided up into a hundred equal pieces. Each piece is one per cent, which can be written as 1%. The term 'per cent' means 'out of a hundred' and is represented by the % symbol.

If we want to describe a part of the whole, we can refer to the number of pieces, or 'per cents', in that part. So, 1% means 'one piece out of a hundred pieces'.

To picture this better, imagine we have a very large bar of chocolate that is divided up into a hundred equal squares.

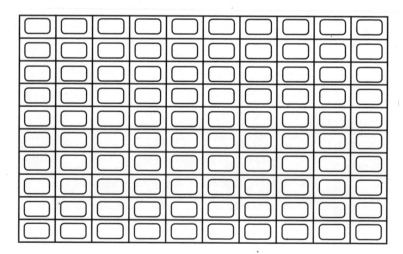

The whole bar of chocolate (100 pieces) would be 100%.

50 pieces would be 'fifty out of a hundred' and could be written as 50%.

If we shade fifty pieces ...

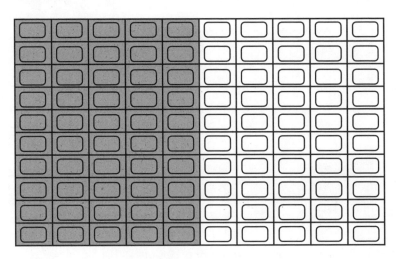

... we can see that 50% is the same as a half ($\frac{1}{2}$).

Similarly, 25% is 25 pieces out of a hundred and is the same as a quarter ($\frac{1}{4}$).

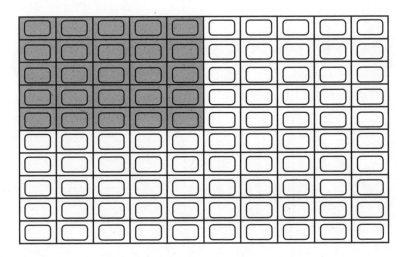

Percentages

75% is 75 pieces out of a hundred and is the same as three quarters ($\frac{3}{4}$).

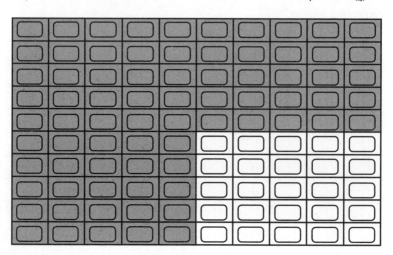

 Remember, the whole bar of chocolate is 100%.

Finding a percentage of a number

Finding a percentage of a number has many practical uses.

With all calculations it is important to know not only *how* to do the calculation, but *why* we are doing it. Otherwise, working with numbers becomes just a series of rules.

The reason for calculating a percentage of a number is to find the value of the part of the number indicated by the percentage. This is easier to understand if we look at an example:

Example

The price of a television set, which normally costs £200, is advertised as being reduced by 25%. What is the reduction in cost? And how much does the television cost now?

To work this out, we need to be able to calculate 25% of the normal price.

In the previous example, the whole bar of chocolate was 100%. In this example, the whole price before any reduction is 100%.

 This is an important idea in percentage work. The whole thing, whether it is a bar of chocolate, or an amount of money, or the length of a piece of wood, etc, is always considered to be 100%.

We can write

100% = £200

Now imagine dividing the £200 up into a hundred pieces. (Imagine sharing £200 between a hundred people if it helps.)

£200 ÷ 100 = £2

So each piece will be £2. 1% of £200 is £2.

Once we know 1% we can find 25% by multiplying the value of 1% by 25.

25% = £2 × 25 = £50

(This can be done manually by the method shown in section 1 of this book, or by using a calculator.)

So, 25% of the price is £50.

In this example, a 25% reduction means a reduction of £50. We would get £50 off the price of the television.

£200 − £50 = £150

Percentages

The sale price would be £150.

This method of finding a percentage of a number can be summed up in two steps:

 Step 1: divide the whole amount of money, length, bar of chocolate, etc, by 100.

 Step 2: multiply the answer to step 1 by the percentage you are looking for.

Practice 1

In this exercise, the whole amount is shown in the first column on the left. Use the two-step method to find firstly 1% of the whole amount, and then the values of the percentages in the other column. The first one has been done for you. If it helps, use a calculator for the dividing and multiplying.

Whole	1%	5%	10%	25%	50%	75%
£300	£300 ÷ 100 = £3	£3 × 5 = £15	£3 × 10 = £30	£3 × 25 = £75	£3 × 50 = £150	£3 × 75 = £225
£500						
£600						
£1000						
£124						
£52						
£250						

In the exercise, all the numbers in the left-hand column represent amounts of money. Remember, they could be numbers representing anything – lengths, numbers of people, weights, and so on.

Here are some other methods of finding a percentage of a number.

Writing the percentage as a fraction

A percentage is a fraction with a denominator of 100, so we can change any percentage into a fraction by putting it over 100. (See the section on page 120 about the relationship between fractions, decimals and percentages.)

Example

Find 5% of £200.

$5\% = \frac{5}{100}$ (five hundredths)

(See page 123 in the section about fractions, percentages and decimals for a full explanation of this.)

5% of £200 $= \frac{5}{100}$ of £200 = £10

So 5% of £200 is £10.

(See the section about finding a fraction of a number on page 50 for the method of doing this.)

Writing the percentage as a decimal

A percentage can be written as a decimal by dividing it by 100.

Example

Find 60% of £350.

$60\% = 0.6$ (nought point six)

(See page 126 in the section on fractions, percentages and decimals for an explanation of this.)

So 60% of £350 is the same as 0.6 of £350

$0.6 \times £350 = £210$

(See the section on multiplying a decimal by a whole number (on page 81) for the method of doing this.)

So 60% of £350 is £210.

Percentages

Using a calculator

 To find a percentage of a number on a calculator, multiply the number by the percentage you want and press the % button.

> **Example**
>
> What is 20% of £750?

Press:

7	5	0	×	2	0	%

Answer on display: 150.

Finding percentages in practical situations

It is often more difficult to carry out a number calculation correctly when the calculation is presented as a written question. The words often obscure the calculation that has to be done. Try the questions in the following exercise. If you can't 'see' the calculation when you first read the question, try underlining or highlighting the numbers, then spend some time thinking about which of the numbers is the percentage and which the whole.

 Remember that the percentage will either have the % sign or the word 'percentage' attached to it.

> **Practice 2**
>
> 1. A washing machine that normally sells for £250 is offered with a 20% reduction in a sale. What is the sale price?
>
> 2. Natalie's gross salary is £25 000 a year. She pays 6% of this in National Insurance. How much National Insurance does she pay?
>
> 3. I invest £500 in a savings account that pays 5% interest per annum (each year). Calculate the interest I will earn in the first year.
>
> 4. A doctor with 240 patients over 60 years old has a flu vaccination target of 75% for the over-sixties. How many patients would have to be vaccinated for the doctor to reach this target?
>
> 5. A family of four spends 30% of its monthly income on rent. If their monthly income is £1700, how much is spent on rent?

Percentage increase and decrease

Sometimes we have to increase or decrease an amount by a particular percentage.

Example

Afzal earns £300 a week. He gets a wage rise of 5%. What is his new wage?

One way of finding Afzal's new wage is by calculating 5% of £300 ...

5% of £300 = (£300 ÷ 100) × 5 = £3 × 5 = £15

... and adding this on to his original wage.

£300 + £15 = £315

Afzal's new wage is £315 a week.

 On some calculators you can do this in one step:

To increase £300 by 5% carry out the calculation for 5% of £300...

... and then press the | **+** | button.

This will add the increase on to the original.

If you were calculating a percentage decrease you would push the | **−** | button at the end of the calculation.

On some calculators the above method does not work. An alternative method is to enter

$$\boxed{3}\ \boxed{0}\ \boxed{0}\ \boxed{+}\ \boxed{5}\ \boxed{\%}$$ for an increase

or $\boxed{3}\ \boxed{0}\ \boxed{0}\ \boxed{-}\ \boxed{5}\ \boxed{\%}$ for a decrease.

Practice 3

1. In 2005 Amina's salary was £26000. In 2006 it rose by 8%. What was her new salary?

2. A company employing 800 people reduced its workforce by 15%. What was the new reduced workforce?

3. In the week before Christmas a supermarket sold 780 packets of mince pies. In the week after Christmas sales of mince pies were down 80%. How many packets were sold in the week after Christmas?

4. When Paul bought a bigger car, his car insurance of £350 a year rose by 5%. What was his new yearly insurance?

5. In 2001 the population of Edinburgh was 5062011. In 2002 it fell by 3%. What was the population in 2002 rounded to the nearest whole number?

Writing one number as a percentage of another

Sometimes it can be useful to work out what one number is as a percentage of another.

Example

A political party would like 60% of its candidates for the next general election to be women. If the party puts up 500 candidates and 300 are women, has the party achieved its 60% target?

In other words:

What percentage is 300 of 500? Is it 60% or more?

Remember that a percentage describes part of a whole. In these types of questions it is important to identify what the 'whole' is.

In this example the 'whole' is the total number of candidates, i.e. 500.

The whole is always 100%, so:

500 = 100%

If 500 = 100%, what percentage is 300?

500 = 100%

300 = ?%

Here are three methods for working out this part of the problem.

Division method

Here, the middle step is finding the percentage for one candidate.

    500 =   100%
    1   =   100% ÷ 500 =   0.2%
    300 =   0.2% × 300 =   60%

The political party has just managed to reach its target of having 60% female candidates.

Short-cut method

Sometimes you might be able to see short cuts which will enable you to go from one number as a percentage to another number as a percentage.

In the example you might be able to 'see' that if 500 = 100%, then 100 must be 20% because it is five times smaller.

    500 =   100%
    100 =   100% ÷ 5   =   20%

If 100 = 20%, then 300 must be 20% × 3 because it is three times bigger.

    500 =   100%
    100 =   100% ÷ 5   =   20%
    300 =   20% × 3   =   60%

In this example, it was relatively easy to go from 500 to 100 because we were dealing with numbers that were both hundreds. Short cuts like this are not always possible – for harder numbers, use the division method.

Fraction method

This is a two-step method:

 Step 1: write the two numbers as a fraction.

Step 2: multiply the fraction by 100 to change it into a percentage. (See the section on page 124 about changing fractions into percentages if you're unsure.)

So in the above example we wanted to write 300 as a percentage of 500.

Step 1: write as a fraction

$\frac{300}{500}$

Step 2: multiply by 100

$\frac{300}{500} \times 100$

This calculation can be done in a number of ways:

$(300 \times 100) \div 500$

or $(300 \div 500) \times 100$

or $(100 \div 500) \times 300$

In all three calculations we are multiplying the 300 and the 100 together and dividing by the 500.

This is exactly what we did in the division method. Both methods are really the same, but the fraction method is slightly quicker.

Increases and decreases as percentages

When writing an increase or a decrease as a percentage, the increase or decrease is always compared with the value of the amount **before** the increase or decrease occurs.

Example

Peter's salary rose from £1200 a month to £1260 a month. What was the percentage increase?

First, we must work out the increase in Peter's salary:

£1260 – £1200 = £60

His initial salary is equivalent to 100%.

£1200   =   100%

£1   =   100% ÷ 1200

£60 =   100% ÷ 1200 × 60   =   5%

Peter's salary rose by 5%.

Practice 4

In the first five questions, write the first number as a percentage of the second:

1. 5   20
2. 18   90
3. 360   2400
4. 80   50
5. 50   50

6. Out of 20 people in an evening class, 12 were women. What percentage of the class were women?

7. A person's salary rose from £20 000 to £20 500. What was the percentage increase?

8. Karla's weekly income is £320. She spends £80 on rent. What percentage of her income does she spend on rent?

9. A dentist has 300 patients. 240 of these are National Health Service (NHS) patients. What percentage are private patients?

10. The price of David's monthly internet connection charge rose from £14.50 a month to £17 a month. What was the percentage rise? Give your answer to the nearest whole number.

Percentages greater than a hundred

We learnt earlier that a whole 'thing' is 100% whether that 'thing' is a bar of chocolate, a sum of money, the number of people in an adult education group or whatever. Sometimes, however, we have to work with percentages that are greater than 100%.

This might seem a strange idea – how can there be more than the whole?

We only have to deal with percentages greater than 100% when we are considering increases in number.

Example

An evening class had 10 students at the beginning of the year. By the end of the first term it had increased by 100%. How many students were in the class at the end of the first term?

To understand what this means, we can think of the starting number as 100%.

100% = 10

So increasing the class by 100% would increase it by 10.

10 + 10 = 20

The class size would rise to 20.

Now let's imagine that by the end of the second term the class size had risen by another 50%.

50% of 10 = 5   (a half of ten equals 5)

20 + 5 = 25

So the class size will have risen to 25.

Over the first two terms the class size has risen 100% and then another 50%. Overall it has risen 150%.

If you are asked to find a percentage greater than 100% of a number, you carry out the calculation in exactly the same way as you would for a percentage less than 100%.

Example

Find 120% of £600.

1%	=	£600 ÷ 100	=	£6
120%	=	£6 × 120	=	£720

(Divide by 100 and multiply by the percentage you require.)

FRACTIONS, DECIMALS AND PERCENTAGES – THE CONNECTION

The equivalence of fractions, decimals and percentages

Fractions, decimals and percentages are all ways of describing parts of a whole. As such, they are all connected and interchangeable.

It is sometimes useful to be able to change from one to another. The following methods explain how this is done.

Fractions to decimals

All fractions can be written as a combination of tenths, hundredths, thousandths, and so on. For example,

$$\frac{1}{2} = \frac{5}{10}$$
$$\frac{1}{4} = \frac{2}{10} + \frac{5}{100}$$
$$\frac{3}{4} = \frac{7}{10} + \frac{5}{100}$$
$$\frac{1}{8} = \frac{1}{10} + \frac{2}{100} + \frac{5}{1000}$$

(You don't need to know how this works, but you can check that it is true by adding the fractions on the right-hand side and cancelling.)

This allows us to put fractions into decimal form:

$$\frac{1}{2} = \frac{5}{10} = 0.5$$
$$\frac{1}{4} = \frac{2}{10} + \frac{5}{100} = 0.25$$

$$\frac{3}{4} = \frac{7}{10} + \frac{5}{100} = 0.75$$

$$\frac{1}{8} = \frac{1}{10} + \frac{2}{100} + \frac{5}{1000} = 0.125$$

In practice, this can be done quickly by dividing the bottom number of the fraction into the top number of the fraction.

$$\frac{1}{2} = \qquad 2\overline{\smash{\big)}\,1.0} \quad \begin{array}{c} 0.5 \end{array}$$

$$\frac{1}{4} = \qquad 4\overline{\smash{\big)}\,1.00} \quad \begin{array}{c} 0.25 \end{array}$$

$$\frac{3}{4} = \qquad 4\overline{\smash{\big)}\,3.00} \quad \begin{array}{c} 0.75 \end{array}$$

$$\frac{1}{8} = \qquad 8\overline{\smash{\big)}\,1.000} \quad \begin{array}{c} 0.125 \end{array}$$

 To change a fraction into a decimal, divide the bottom number into the top

Notice how a decimal point and one or more zeros are added to the top number to complete the division. (The carrying has not been shown in these examples – have a look at the section on page 85 about dividing decimals if you're not sure. You can always do these divisions on a calculator if you need to.)

Mixed fractions to decimals

If the fraction has a whole number part, change the fraction part as shown above and include the whole number part in your answer.

$$2\frac{3}{4} = 2.75$$

Fractions, decimals and percentages

Change the following fractions into decimals:

1. $\frac{3}{8}$

2. $\frac{7}{10}$

3. $\frac{4}{5}$

4. $\frac{1}{20}$

5. $2\frac{1}{4}$

Decimals to fractions

We can use the reverse of the above process to change decimals into fractions.

i If a decimal has one decimal place it can be written as tenths.

i If a decimal has two decimal places it can be written as hundredths.

i If a decimal has three decimal places it can be written as thousandths.

$0.3 = \frac{3}{10}$

$0.19 = \frac{19}{100}$

$0.127 = \frac{127}{1000}$

Sometimes the fraction can be cancelled down:

$0.2 = \frac{2}{10} = \frac{1}{5}$ (dividing top and bottom by 2)

$0.15 = \frac{15}{100} = \frac{3}{20}$ (dividing top and bottom by 5)

$0.284 = \frac{284}{1000} = \frac{142}{500}$ (dividing top and bottom by 2) $= \frac{71}{250}$ (dividing by 2 again)

To change a decimal to a fraction put over 10, 100, 1000 ... and cancel down if possible. The bottom number of the fraction should have as many zeros as there are decimal places in the decimal.

Decimals to mixed fractions

If the decimal has a whole number part, change the decimal part as above and include the whole number part in your answer.

$$4.5 = 4\tfrac{1}{2}$$

Practice 2

Change the following decimals into fractions in their lowest terms:

1. 0.4
2. 0.26
3. 0.124
4. 4.6
5. 12.14

Percentages to fractions

A percentage is a fraction with a denominator (bottom number) of a hundred. To change a percentage to a fraction, put the percentage over 100 and cancel down if possible.

$25\% = \frac{25}{100} = \frac{5}{20}$ (cancelling by 5) $= \frac{1}{4}$ (cancelling by 5)

$50\% = \frac{50}{100} = \frac{5}{10}$ (cancelling by 10) $= \frac{1}{2}$ (cancelling by 5)

$75\% = \frac{75}{100} = \frac{15}{20}$ (cancelling by 5) $= \frac{3}{4}$ (cancelling by 5)

The fraction equivalent of the above percentages may be familiar to you as they are often used in everyday life. However, the above method may be used with any percentage.

Example

What is 38% written as the simplest possible fraction?

$$38\% = \frac{38}{100} = \frac{19}{50} \text{ (cancelling by 2)}$$

To change a percentage to a fraction, put the percentage over a hundred. Then cancel it down if you can.

Percentages greater than 100%

The above method can be used to change a percentage greater than 100% into a fraction. The fraction obtained can be written as a mixed fraction.

Example

What is 250% as a fraction?

$250\% = \frac{250}{100} = \frac{25}{10}$ (cancelling by 10) $= \frac{5}{2}$ (cancelling by 5) $= 2\frac{1}{2}$

Practice 3

Change the following percentages into fractions in their lowest terms:

1. 35%
2. 82%
3. 46%
4. 8%
5. 120%

Fractions to percentages

To change a fraction to a percentage, we have to change the denominator (bottom number) of the fraction to a hundred. We then change the numerator (top number) accordingly. (See the section on page 43 about equivalent fractions.) The new top number tells us the percentage.

$\frac{4}{10} = \frac{40}{100}$ (multiplying top and bottom of the fraction by 10) $= 40\%$

$\frac{3}{5} = \frac{60}{100}$ (multiplying top and bottom of the fraction by 20) $= 60\%$

$\frac{3}{20} = \frac{15}{100}$ (multiplying top and bottom of the fraction by 5) $= 15\%$

In all of the examples so far it has been fairly straightforward to change the bottom number into a hundred. With most fractions it is more difficult. The following is a quick way of converting any fraction to a percentage:

To change a fraction into a percentage, multiply the top number by a hundred and then divide by the bottom number.

$\frac{4}{10}$ → multiplying the top number by 100 → $\frac{4}{10} \times 100 \% = \frac{400}{10}\% = 40\%$

$\frac{3}{5}$ → multiplying the top number by 100 → $\frac{3}{5} \times 100 \% = \frac{300}{5}\% = 60\%$

$\frac{3}{20}$ → multiplying the top number by 100 → $\frac{3}{20} \times 100 \% = \frac{300}{20}\% = 15\%$

Changing a mixed fraction into a percentage

To change a mixed fraction into a percentage, make the fraction top-heavy (the method for this is on page 54 in the section about fractions) and multiply by a hundred as above.

Example

What is $1\frac{1}{2}$ as a percentage?

$1\frac{1}{2} = \frac{3}{5}$ → multiplying the top number by 100 → $\frac{3}{2} \times 100 \% = \frac{300}{2}\% = 150\%$

Practice 4

Change the following fractions into percentages:

1. $\frac{7}{20}$

2. $\frac{2}{5}$

3. $\frac{3}{8}$

4. $\frac{15}{50}$

5. $1\frac{3}{4}$

Fractions, decimals and percentages

Percentages to decimals

A percentage can be written as a fraction by putting it over a hundred. (See the section on page 123 about changing percentages to fractions.)

A fraction can be changed into a decimal by dividing the bottom number into the top. (See the section on page 120 about changing fractions to decimals.)

It follows that a percentage can be changed into a decimal by dividing it by 100.

Example

What is 26% as a decimal?

26% can be written as a fraction by putting it over 100:

$$26\% = \frac{26}{100}$$

$\frac{26}{100}$ can be written as a decimal by dividing the bottom number into the top:

$$\frac{26}{100} = 26 \div 100 = 0.26$$

This can be done on a calculator, or on paper as a long division, or by putting in a decimal point two places back from the end of the number (see the short cut for dividing by 100 on page 87 in the decimals section).

$$26\% = 0.26$$

In practice, it is unnecessary to show these three separate stages.

Example

Change 59% into a decimal.

$$59 \div 100 = 0.59$$

 To change a percentage to a decimal, divide the percentage by 100.

Percentages greater than 100%

The same method can be used to change a percentage greater than 100% into a decimal. The decimal number obtained will have a whole number part.

Example

Change 175% into a decimal.

$175 \div 100 = 1.75$

Practice 5

Change the following percentages into decimals:

1. 62%
2. 555%
3. 2%
4. 18%
5. 142%

Decimals to percentages

Changing a decimal into a percentage is the reverse of the process for changing a percentage into a decimal.

 To change a decimal to a percentage, multiply by 100.

Example

Change 0.45 into a percentage.

$0.45 \times 100 = 45\%$

This can be done on a calculator, or on paper as a long multiplication, or by moving the decimal point two places to the right (see the short cut for multiplying by 100 on page 82 in the decimals section).

Practice 6

Change the following decimals into percentages:

1. 0.68
2. 0.36
3. 0.16
4. 0.02
5. 1.05

Converting anything and everything

If you are someone who finds diagrams helpful, this one might be useful. It shows all of the above methods in one diagram.

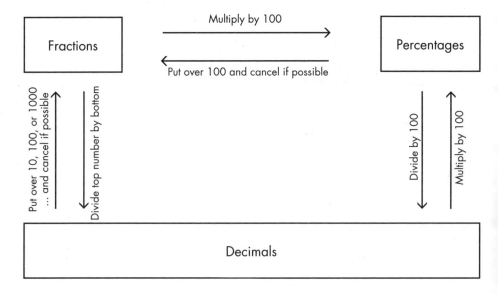

Some people find this sort of diagram confusing and unhelpful. Don't worry if you do, there are other ways of remembering the methods we have discussed in this section:

Make up your own diagram using symbols or words that will help you remember when to divide and when to multiply.

Write out the methods in your own words – people often find that doing this makes the methods clearer and easier to remember.

You don't have to memorize the methods, because you can always look them up when you need them. That is what this book is for! Repeated use will help you eventually remember the method.

Most important, and most useful of all, a thorough understanding of what fractions, percentages, and decimals are will help when converting between them. For example, if you know that a percentage is just a fraction with a denominator (bottom number) of a hundred, then changing a percentage to a fraction is simply a matter of writing it with a denominator of a hundred. Look back over the sections on fractions, percentages and decimals to refresh your understanding.

RATIO AND PROPORTION

Ratio basics

A **ratio** is used to compare one quantity with another.

> ### Example
>
> Rees earns £300. Davina earns £400 a week. What is the ratio of their wages?

Comparing their weekly wages as a ratio:

Rees Davina

300 : 400

We say that the ratio of Rees's wage compared to Davina's is 300 to 400. The **:** sign between the two quantities is the **ratio symbol**.

When comparing amounts using a ratio we don't need to specify the units (the £ signs in this example) as we are comparing the size of the amounts.

 To compare two amounts as a ratio, write the two numbers side by side with the ratio symbol between them.

This idea can be extended to three, or four, or as many numbers as we wish, though it is more usual to compare only two or three numbers.

> ### Example
>
> Asha is 120 centimetres tall, Ruth is 150 centimetres tall and Jamil is 180 centimetres tall. What is the ratio of their heights?

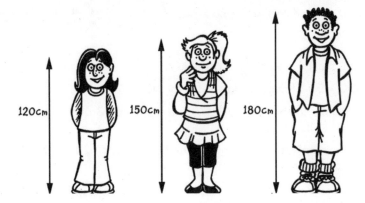

Comparing their heights as a ratio:

120 : 150 : 180 – notice that the ratio symbol is placed between each number

 Quantities have to be in the same units before they can be compared using a ratio.

Example

Billy has £1 to spend on sweets. Mary has 50p to spend. Write down the amounts of money they have as a ratio.

We can't write these amounts as a ratio straight away, because we can't compare pounds with pence. Before writing them as a ratio, the amounts must be written in the same units.

Changing Billy's £1 into pennies, Billy has 100p to spend. Billy's money is now in the same units as Mary's and we can write the two amounts as a ratio.

100 : 50

Simplifying ratios

Ratios can be thought of as a type of fraction. Like a fraction, a ratio can be cancelled to its lowest or simplest form.

To cancel, we find a whole number which will divide exactly into both sides of the ratio. (See the section about cancelling fractions on page 46 if you're not sure about this.)

In the first example, the ratio of Rees's wage to Davina's wage was 300 : 400. This can be cancelled as follows:

300 : 400 = 30 : 40 (dividing both sides by 10) = 3 : 4 (dividing both sides by 10 again)

The ratio can also be cancelled in a single step:

300 : 400 = 3 : 4 (dividing both sides by 100)

There is no whole number that will divide exactly into 3 and 4, so the ratio has been cancelled to its lowest form.

In the second example, the ratio of the three heights was 120 : 150 : 180. This can be cancelled as follows:

120 : 150 : 180 = 12 : 15 : 18 (dividing by 10) = 4 : 5 : 6 (dividing by 3)

There is no whole number which will divide exactly into 4, 5 and 6, so the ratio has been cancelled to its lowest form.

 To cancel a ratio, divide all the numbers by the same amount each time.

Practice 1

Cancel each of the following ratios to its lowest form:

1. 2 : 8
2. 16 : 48
3. 50 : 200
4. 18 : 24 : 60
5. 125 : 500 : 1000

Fractional ratios

Sometimes we have to work with ratios that contain fractions.

Example

A recipe requires 2 kilograms of flour to be mixed with $\frac{1}{2}$ kilogram of sugar. What is the ratio of flour to sugar?

The ratio of flour to sugar is $2 : \frac{1}{2}$.

However, ratios are usually written as whole numbers. There are different ways to simplify fractional ratios, depending on how many of the numbers are fractions.

Simplifying ratios when one number is a fraction

When cancelling a ratio containing a fraction and a whole number to its lowest form, the rule is to *multiply* the fractional part of the ratio up to the nearest whole number. The whole number part of the ratio is then multiplied up the same number of times.

Example

Write the ratio $2 : \frac{3}{4}$ in its lowest form.

The first step is to multiply $\frac{3}{4}$ up to the lowest possible whole number. The easiest way to do this is by multiplying the top number of the fraction – the 3 – by the bottom number – the 4. This will change the fraction into a top-heavy fraction, which can then be divided out to give the lowest possible whole number.

$$\frac{3}{4} \times 4 = \frac{12}{4} = 12 \div 4 = 3$$

The next step is to multiply the whole number part of the ratio by the same number as we used to multiply the fraction:

$$2 \times 4 = 8$$

We can now write the ratio in its simplest form:

$$2 : \frac{3}{4} = 8 : 3$$

You might be able to see from looking at these two ratios that the quickest way of simplifying a ratio like this is to multiply both sides of the ratio by the bottom number of the fraction.

$$2 : \frac{3}{4} = 2 \times 4 : \frac{3}{4} \times 4 = 8 : 3$$

 The above methods of making a fractional ratio up to the nearest whole number only work if the fraction is in its simplest form to begin with. If it isn't then you must simplify it first.

Ratio and proportion

Example

What is $6:\frac{4}{16}$ in its simplest form?

$6:\frac{4}{16}$ cancelling the fraction to its simplest form $6:\frac{1}{4}$

Now we can write the ratio in its simplest form by multiplying each side of the ratio by 4.

$6 \times 4 : \frac{1}{4} \times 4 = 24:1$

To write a ratio containing a whole number and a fraction in its simplest form, first make sure the fraction is in its simplest form. Then multiply the whole number and the top number by the bottom number of the fraction.

Practice 2

Cancel each of the following ratios to its simplest form:

1. $2:\frac{1}{2}$

2. $5:\frac{1}{4}$

3. $3:\frac{3}{4}$

4. $1:\frac{1}{3}$

5. $10:\frac{3}{8}$

Simplifying ratios when both numbers are fractions

When cancelling a ratio containing two or more fractions, the parts of the ratio must be multiplied by the lowest common denominator of the fractions. (See the section on page 60 about finding the lowest common denominator of fractions if you're not sure how to do this.)

Example

A cake recipe requires $\frac{1}{2}$ kilogram of sugar and $\frac{1}{4}$ kilogram of butter. The ratio of sugar to butter is $\frac{1}{2}:\frac{1}{4}$. What is the simplest form of this ratio?

To write this in its simplest form, we multiply by the lowest common denominator of 2 and 4.

The lowest common denominator is the smallest number that can be divided exactly by both numbers. For 2 and 4 this would be 4.

Multiplying both fractional parts by 4:

$$\tfrac{1}{2} \times 4 : \tfrac{1}{4} \times 4 = 2 : 1$$

Ratios that include mixed fractions

If one or more of the parts of a ratio is a mixed fraction, change the mixed fraction to a top-heavy fraction (see the section on page 53 about mixed and top-heavy fractions) and then proceed as above.

Example

Write the ratio $1\tfrac{1}{2} : \tfrac{1}{4}$ in its lowest form.

$$1\tfrac{1}{2} : \tfrac{1}{4} = \tfrac{3}{2} : \tfrac{1}{4}$$

The lowest common denominator is 4, so multiply both parts of the ratio by 4:

$$\tfrac{3}{2} \times 4 : \tfrac{1}{4} \times 4 = \tfrac{12}{2} : \tfrac{4}{4} = 6 : 1$$

Practice 3

Write the following ratios in their simplest form:

1. $\tfrac{1}{5} : \tfrac{1}{2}$

2. $\tfrac{2}{3} : \tfrac{1}{4}$

3. $\tfrac{1}{8} : \tfrac{1}{3}$

4. $1\tfrac{1}{2} : \tfrac{1}{4}$

5. $2\tfrac{1}{2} : \tfrac{2}{3}$

Proportion

Ratio is used to **compare two or more quantities**. A related word, **proportion**, is used to express one part as a **fraction of the whole**.

Ratio and proportion

In the very first example in this chapter, Davina earned £400 a week and Rees earned £300. We saw that the ratio of Rees's earnings to Davina's earnings was $300:400$ (which can be cancelled down to $3:4$). Now we can work out each person's proportion of their total earnings:

The total earned by both Davina and Rees is £400 + £300 = £700.

Davina's wage as a *proportion* of the whole is $\frac{400}{700}$ (400 *out of* 700).

This can be cancelled down in the usual way:

$\frac{400}{700} = \frac{40}{70}$ (cancelling by 10) $= \frac{4}{7}$ (cancelling by 10)

Rees's wage as a *proportion* of the whole is $\frac{300}{700}$ (300 *out of* 700).

$\frac{300}{700} = \frac{30}{70}$ (cancelling by 10) $= \frac{3}{7}$ (cancelling by 10)

We have seen that we can only compare two or more quantities as a ratio if the units of all the quantities are the same. The same rule applies when writing one part as a proportion of the whole.

Example

Theresa is mixing orange squash for her daughter's party. To every litre of water she adds 250 millilitres of juice. What is the proportion of juice to water?

The total amount of juice is found by adding 1 litre to 250 millilitres. Here the units are different, so we need to change the litre into millilitres before adding. 1 litre = 1000 millilitres (see section on measurement).

First we must work out the total amount of squash:

1000 millilitres + 250 millilitres = 1250 millilitres

Then we can calculate the proportion of juice:

$\frac{250}{1250} = \frac{25}{125}$ (cancelling by 10) $= \frac{5}{25}$ (cancelling by 5) $= \frac{1}{5}$ (cancelling by 5)

The proportion of juice in the orange squash is $\frac{1}{5}$.

Proportional parts

Sometimes we need to divide a quantity into two or more parts so that the proportions of the parts are in a given ratio.

Example

Two friends share a lottery win of £2 450 240 but, because they put in different amounts to buy the lottery tickets they don't share it equally. Instead, it is shared in the ratio 3 : 2. How much does each win?

In questions like this it is useful to think in terms of 'shares'. So in this example, one friend will get 3 shares while the other will get 2 shares.

The total number of shares is found by adding together the two parts of the ratio.

Total number of shares = 3 + 2 = 5

The total of the shares is equal to the total amount to be divided.

5 shares = £2 450 240

Now we can work out the value of one share:

If 5 shares = £2 450 240, then 1 share will be £2 450 240 ÷ 5 = £490 048.

Finally, we can work out the amounts won by each friend:

If   1 share   =   £490 048
then 3 shares   =   £490 048 × 3 = £1 470 144
and   2 shares   =   £490 048 × 2 = £980 096

One friend gets £1 470 144 and the other gets £980 096.

In these questions you can always check that your calculations are correct by adding the two parts together. They should come to the same total as was originally divided.

£1 470 144 + £980 096 = £2 450 240

Ratio and proportion

In this example, the use of the word 'share' was appropriate as it involved two friends sharing a lottery win. However, even in situations where there are no obvious shares involved, it is is often useful to think in terms of shares.

Example

A cook mixes flour, sugar and butter in the ratio $5:2:1$ to make 2400 grams of cake mixture. What weight of each ingredient does the cook need?

Here we are not dividing up an amount of money but it is still useful to think in terms of shares: flour has 5 shares, sugar has 2 shares and butter has 1 share.

Total shares $= 5 + 2 + 1 = 8$.

1 share	=	2400 grams	÷	8	=	300 grams
5 shares	=	300 grams	×	5	=	1500 grams
2 shares	=	300 grams	×	2	=	600 grams
1 share	=	300 grams	×	1	=	300 grams

The cook will need to use 1500 grams of flour, 600 grams of sugar and 300 grams of butter.

Checking the total: $1500 + 600 + 300 = 2400$, which suggests that the calculations are correct.

Practice 4

1. Divide £5000 in the ratio $3:2$.
2. Divide £81 000 in the ratio $2:3:4$.
3. Divide 3592 grams in the ratio $1:3$.
4. Divide 260 centimetres in the ratio $3:\frac{1}{4}$ (hint: put the ratio in its simplest form first).
5. Divide £490 in the ratio $1:\frac{1}{2}:\frac{1}{4}$.

Sometimes, when a quantity is divided in a given ratio, we know the value of one of the parts and need to find the other.

To make mortar a builder mixes cement and sand in the ratio 2:5. He has 6 kilograms of cement. How much sand will he need?

In the previous example, we knew the total amount of cake mixture, so we added the two parts of the ratio together to enable us to find one share.

In this example, we know the value of one part of the ratio – the amount of cement the builder has. We use this to find one share.

2 shares	=	6 kilograms		
1 share	=	6 kilograms ÷ 2	=	3 kilograms
5 shares	=	3 kilograms × 5	=	15 kilograms

The builder needs 15 kilograms of sand.

Practice 5

1. Two friends share £1800 in the ratio 5:4. How much does each receive?

2. An adult education group has 20 students. The ratio of women to men is 3:2. How many women are in the group?

3. 1000 people were surveyed and asked whether they were in full-time employment, part-time employment, or unemployed. The ratio of full-time workers, part-time workers and unemployed people was 5:3:2. How many people were in each category?

4. The ratio of Welsh, Scottish, and English holidaymakers on a package holiday to Spain was 3:2:1. If 18 Welsh people were on the holiday, how many Scottish and English people were there?

5. A lottery grant was shared between charity A and charity B in the ratio 3:4. If charity A received £42000, how much did charity B receive?

Direct Proportion

Two or more quantities are said to be in **direct proportion** if they both **increase or decrease in the same ratio**.

If	1 kilogram of oranges costs £1.50
then	2 kilograms of oranges cost £3
and	3 kilograms of oranges cost £4.50
and so on …	

Here the two variables we are considering are the *weight* of the oranges and the *cost* of the oranges.

Weight	Cost
1 kilogram	£1.50
2 kilograms	£3
3 kilograms	£4.50

The weight and the cost are in direct proportion because as the weight increases, the price also increases *in the same ratio*.

Or put more simply, if the weight doubles then the cost doubles, if the weight increases threefold then the cost increases threefold, and so on. If the weight increases tenfold (if we bought 10 kilograms of oranges) then the cost would also increase tenfold (the cost would be £15). *Both* would be ten times more.

Have a look at another case where you might encounter direct proportion:

The following table shows the cost of renting a car.

Number of days	Cost
1	£30
2	£60
3	£90
4	£120
5	£150
6	£180
7	£200

Look carefully at the number of days and the costs. For the first 6 days, the number of days and the cost are in direct proportion – double the number of days and the cost doubles, treble the number of days and the cost trebles … multiply the number of days by 6 and the cost gets multiplied by 6. For the first 6 days, the number of days and the cost increase *in the same ratio*.

This pattern breaks down on the seventh day (perhaps because there is a discount for hiring for a full week). After the sixth day the cost still goes up but not in the same ratio as the number of days. The relationship between number of days and cost is no longer directly proportional.

Both these examples illustrate situations that may be familiar. It might seem that we are stating the obvious. *Of course* the cost of 2 kilograms of oranges is double the cost of 1 kilogram, or the cost of hiring a car for 3 days is three times more than hiring a car for 1 day. But even though these examples might seem obvious, they illustrate a very important and useful idea in numeracy:

 If two or more things (variables) are in direct proportion, then knowing one set of values for the variables allows us to find other sets of values.

Example

John is organizing a meal in a restaurant for a group of friends. Originally 7 people were going to eat and the price quoted by the restaurant was £84. What would be the price if only 5 could go?

In this example, we know the price for 7 people and we want to find the price for 5 people. We want to go from one value that we know (the cost for 7 people) to a value that we don't know (the cost for 5 people).

7 people cost £84
5 people cost **?**

It is difficult to go directly from the cost for 7 people to the cost for 5 people, so as an intermediate step we find the cost for 1 person.

7 people cost   £84
1 person costs   £84 ÷ 7 (seven times less)   =   £12
5 people cost   £12 × 5 (5 times more than 1 person)   =   £60

The cost for 5 people is £60.

This three-step process is useful in many different contexts.

Example

A recipe asks for 120 grams of sugar to be mixed with 720 grams of flour. How much flour would be needed for 500 grams of sugar?

Using the three-step process shown above:

120 grams of sugar requires 720 grams of flour

1 gram of sugar requires $720 \div 120$ grams of flour $= 6$ grams of flour

500 gram of sugar requires 6×500 grams of flour $= 3000$ grams of flour

3000 grams (or 3 kilograms) of flour is needed.

Inverse proportion

Two or more variables are said to be in inverse proportion if an increase in one produces a decrease in the other in the same ratio, or if a decrease in one produces an increase in the other in the same ratio.

Example

Two men working at the same rate take 3 days to build a wall. How long would it take three men working at the same rate?

2 men take 3 days

1 man takes ? days

Be careful: it might not be obvious that that the question involves *inverse* proportion. In questions involving direct proportion, the middle step – finding the value for 1 – requires a division to be carried out. (Look back at the examples involving direct proportion.). If we used division in the second step in this example we would get the very unrealistic answer that 1 man would take $3 \div 2$ days $= 1\frac{1}{2}$ days to build the wall. In other words, one man working alone would build the wall in less time than 2 men working together!

Remember that, with inverse proportion, as one value decreases the other *increases*. So, as the number of men decreases, the number of days will increase.

In problems involving inverse proportion the middle step requires a multiplication and not a division.

2 men take 3 days

1 man takes 3×2 days $= 6$ days

For the final step – finding out how long it would take 3 men to build the wall – we must again remember that, in inverse proportion, as one value increases the other decreases. So as the number of men now increases, the number of days will *decrease*. The third step requires a division, rather than a multiplication as in direct proportion problems.

2 men take 3 days

1 man takes 3 × 2 days = 6 days

3 men take 6 ÷ 3 days = 2 days

3 men would take two days to build the wall.

 If you are worried about multiplying when you should divide or dividing when you should multiply, ask yourself the question – 'does my answer make sense?' Often, this simple check will identify if you have done the wrong thing.

Practice 6

1. 7 metres of curtain material costs £14. How much would 9 metres cost?

2. A pasta recipe for 5 people requires 375 grams of pasta. How much pasta would be required for 7 people?

3. Renting a car for 4 days costs £160. How much would it cost to rent the car for 10 days?

4. An athlete running at a steady pace covers 2 miles in 20 minutes. How long would it take the same athlete, running at the same rate, to cover 9 miles?

5. 3 women, working at the same rate, take 7 hours to decorate a house. How long would it take 2 women working at the same rate?

Applications of ratio and proportion

We will look at three examples of the use of ratio and proportion.

Maps

Most maps are drawn to a given scale. The scale is usually shown as a ratio such as 1 : 25 000, 1 : 50 000 or 1 : 100 000, depending on how detailed the map is.

The scale shows how much bigger distances in real life are than distances on the map.

SCALE 1:250,000

Glasgow · · Edinburgh
30cm

A road atlas of the British Isles has a scale of 1 : 250 000. On the map the distance between Edinburgh and Glasgow is shown as 30 centimetres. What would be the real-life distance?

The scale of the map tells us that real-life distances are 250 000 times bigger than map distances, so the real-life distance between Edinburgh and Glasgow is:

30 centimetres × 250 000 = 7 500 000 centimetres

Seven and a half million centimetres? This is the correct distance, but it would be unusual to describe the distance between two towns in centimetres. Instead, we can change the distance into metres by dividing by 100 (there are 100 centimetres in a metre – see page 152 in the section about measurement).

7 500 000 centimetres ÷ 100 = 75 000 metres

We can then change the distance in metres into kilometres by dividing by 1000 (there are 1000 metres in a kilometre – again, see page 152 in the section about measurement).

75 000 metres ÷ 1000 = 75 kilometres

The real-life distance between the two cities is 75 kilometres.

 To change distances on the map into real-life distance, multiply by the second part of the ratio. Then change the answer to the appropriate units.

Distances in real life can also be changed into map distances if we know the scale of the map, as shown in the next example.

Example

The distance between Newcastle and Edinburgh is 170 kilometres. What would be the distance on the road atlas of the British Isles which has a scale of 1 : 250 000?

The scale of the map tells us that map distances are 250 000 times smaller than real-life distances, so the real-life distance between Newcastle and London is:

170 kilometres ÷ 250 000

Doing this division on a calculator gives an answer of 0.00068 kilometres.

This is correct, but distances on maps are usually described in centimetres. As a first step, we can change the answer into metres by multiplying by 1000 (there are 1000 metres in a kilometre – see the section about measurement on page 152).

0.00068 kilometres × 1000 = 0.68 metres

We can then change the distance from metres into centimetres by multiplying by 100 (there are 100 centimetres in a metre).

0.68 metres × 100 = 68 centimetres

The distance between Newcastle and London in the road atlas would be 68 centimetres.

To change real-life distances into distances on the map, divide by the second part of the ratio. Then change the answer to the appropriate units.

Model making

The dimensions (length, width and height) of a **scale model** of an object are in the same proportions as the dimensions of the original full-size object. For example, if the length of the full-size object is twice its width then the length of the model will be twice *its* width.

In the following example the dimensions of the model tractor have been scaled down by a factor of ten. We can say it has been scaled down in the ratio 1:10.

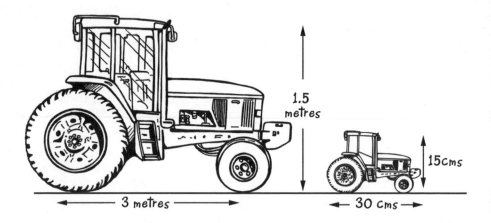

Notice that the length and height of the real tractor have been reduced by a factor of ten in the model.

Length of real tractor	=	3 metres		
Length of model	=	3 metres ÷ 10	=	30 centimetres
Height of real tractor	=	1.5 metres		
Height of model	=	1.5 metres ÷ 10	=	15 centimetres

Notice, also, that the proportions of length and height in the real and model tractor are the same. (Length is double the height in both.)

If we know the scale ratio of a model and the dimensions of the real-life object then we can find the dimensions of the model.

Example

Roger is making a 1 : 30 scale model of the steam engine, the Flying Scotsman. The actual length of the engine is 21 metres. What length should he make the model?

The length of the model is 30 times smaller than the length of the actual engine. To find the length of the model we have to divide 21 metres by 30:

21 metres ÷ 30 = 0.7 metres or 70 centimetres.

Roger should make his model 70 centimetres long.

We can also find the dimensions of a real-life object if we know the dimensions of a model and the scale ratio.

Example

A 1 : 200 scale model of Blackpool tower is 78 centimetres high. What is the height of the real tower?

The height of the real tower will be 200 times bigger than the model.

78 centimetres × 200 = 15 600 centimetres

Dividing by 100 to change to metres:

15 600 centimetres ÷ 100 = 156 metres

Blackpool tower is 156 metres high.

Scale drawings

Construction projects usually use scale drawings to represent what is to be built.

As in map making and modelling, a suitable ratio is used to scale down the dimensions of the project so that a drawing can be made. Again, if the ratio being

Ratio and proportion

used is known, then dimensions in real life can be changed into dimensions on the plan and vice versa.

Example

The length of the living room on house plans, drawn to a scale of 1 : 200, is 5 centimetres. What is the actual length of the living room?

The actual length will be 200 times the length in the plans.

5 centimetres × 200 = 1000 centimetres = 10 metres (dividing by 100 to convert to metres)

The living room is 10 metres long.

Practice 7

1. The distance between two towns on a map with a scale of 1 : 200 000 is 7 centimetres. What is the real distance between these towns? Give your answer in kilometres.

2. The distance between Edinburgh and Fort William is 220 kilometres. What would the distance be on a map with a scale of 1 : 750 000 ? Give your answer to the nearest centimetre.

3. A model car, built to the scale of 1 : 10, is 32 centimetres long. What is the real length of the car in metres?

4. Richard is making a model of a Viking longboat to the scale of 1 : 50. The actual length of the boat is 20 metres. What length should he make the model?

5. The plans of a new shopping centre show the car park as being 15 centimetres long. If the plans are drawn to a scale of 1 : 500, what will be the actual length of the car park? Give your answer in metres.

MEASUREMENT

Systems of measurement

This part of the book is about the measurement of length, weight and capacity. (The measurement of time and temperature will be dealt with later in the book.)

At present in Britain, two systems of measurement are in use: the **imperial system** and the **metric system**.

The imperial system is the 'old' system with which many of us are still familiar. For example, British road signs give distance in miles.

This is being replaced by the metric system, although both are still in use. All schools now teach measurement using only the metric system, although some schools teach the relationship between the two systems and how to change from one to another.

We are going to look mainly at the metric system of measurement. In the appendix at the back of the book you will find a conversion table which will show you the connection between the units in both systems.

The advantage that the metric system has over the imperial system, and one reason why it is gradually replacing it, is its simplicity and 'sameness'. The structure of the system is the same for measurement of length, weight or capacity. In the imperial system, the units for length, weight and capacity are all different, each set of units having developed haphazardly over centuries. The metric system has been designed to make things less complicated. For those raised on the imperial system it might seem hard to believe, but it's true – once you learn how it works, the metric system is easier to use.

How the metric system works

In each of the three fields of measurement – length, weight and capacity – there is one main unit. All other units in that field are compared to that main unit.

Main unit of length

The main unit of length is the **metre**. This is a distance slightly longer than the old imperial yard, roughly the distance of an adult's stride.

Main unit of weight

The main unit of weight is the **gram** or **gramme**. This is roughly the weight of a paper tissue or matchstick.

The property that is known as 'weight' in everyday language is strictly called mass. Scientists would tell you that they use grams as a measure of mass, not weight. In this book, though, we will stick to the common usage and refer to measurements in grams using the word 'weight'.

Main unit of capacity

Capacity is a measure of how much something holds. It is also sometimes called volume. The main unit of capacity is the **litre**. A standard carton of fruit juice usually contains a litre.

Other units

Now comes the clever part:

 The other metric units used for measuring length, weight and capacity are either a hundred or a thousand times bigger or smaller than the main unit.

For measuring length, the other units are:

millimetre	centimetre	**metre**	kilometre

For measuring weight, the other units are:

milligram	*centigram**	**gram**	kilogram

For measuring capacity, the other units are:

millilitre	centilitre	**litre**	*kilolitre**

*These are rarely used in practice

You will notice that all the units of length have **metre** in their name, all the units of weight have **gram** in their name, and all the units of capacity have **litre** in their name.

You will also notice that the parts of the names that go before metre, gram, or litre are the same in each field of measurement: **milli-**, **centi-** and **kilo-**.

Milli- means a **thousandth**
Centi- means a **hundredth**
Kilo- means a **thousand**

A **millimetre** is a **thousandth** of a **metre**
A **centimetre** is a **hundredth** of a **metre**
A **kilometre** is a **thousand metres**

A **milligram** is a **thousandth** of a **gram**
A **centigram** is a **hundredth** of a **gram**
A **kilogram** is a **thousand grams**

A **millilitre** is a **thousandth** of a **litre**
A **centilitre** is a **hundredth** of a **litre**
A **kilolitre** is a **thousand litres**

Abbreviations

The following abbreviations are commonly used:

millimetres	**mm**	milligrams	**mg**	millilitres	**mL**
centimetres	**cm**	centigrams	**cg**	centilitres	**cL**
metres	**m**	grams	**g**	litres	**L**
kilometres	**km**	kilograms	**kg**	kilolitres	**kL**

Notice that that a capital L is often used for the abbreviation of litres to avoid confusion with the number 1, especially when litres are being written on their own. However, a small l is also used, especially in the abbreviation **ml** for millilitres.

Notice also that two of the units – centigrams and kilolitres – are in grey. This is to draw your attention to the fact that these units are rarely used. However, they have been included throughout the section on measurement to show that the structure of the metric system is the same for measuring length, weight and capacity.

Converting metric units

When working with the metric system we often have to convert from one unit to another.

Example

Andrea wants to buy a curtain rail. On measuring her window she finds that she will need a rail 1.75 metres long. However, when she goes to buy one she finds that the lengths of the curtain rails in the shop are in millimetres. How many millimetres long is her curtain rail?

To answer this question, we need to change 1.75 metres into millimetres.

Before looking at how we can do this, let's look again at the connection between units of length.

Instead of writing:

A millimetre is a thousandth of a metre
A centimetre is a hundredth of a metre
A kilometre is a thousand metres

we can write:

1 metre	=	1000 millimetres
1 metre	=	100 centimetres
1 kilometre	=	1000 metres

If 1 metre = 1000 millimetres, then:

2 metres = 2000 millimetres (doubling the 1000 millimetres or multiplying 1000 millimetres by 2)

3 metres = 3000 millimetres (1000 millimetres multiplied by 3)

7 metres = 1000 millimetres × 7 = 7000 millimetres

 To change metres into millimetres, multiply by 1000.

So we can change 1.75 metres into millimetres by multiplying by 1000.

1.75 metres × 1000 = 1750 millimetres

Andrea needs to buy a curtain rail 1750 millimetres long.

Conversion factors

What if Andrea had measured her window in centimetres? How could she change centimetres into millimetres?

To help answer this question look at a ruler (not drawn to scale). If your ruler is very old it might be only marked off in inches, but most rulers show centimetres and millimetres. Millimetres are the small divisions between the centimetres.

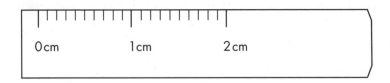

Count the number of millimetres in one centimetre. You should find that there are 10 millimetres in a centimetre.

10 millimetres	=	1 centimetre
100 centimetres	=	1 metre
1000 metres	=	1 kilometre

These are the **conversion factors** for metric length. They tell you how many of one unit there are in another. The conversion factors for weight and capacity are similarly:

10 milligrams	=	1 centigram
100 centigrams	=	1 gram
1000 grams	=	1 kilogram
10 millilitres	=	1 centilitre
100 centilitre	=	1 litre
1000 litres	=	1 kilolitre

If Andrea had measured her window in centimetres, she would have gone to the shop with a measurement of 175 centimetres. She would have wanted to change these into millimetres. From above, we know that 10 millimetres = 1 centimetre, so

175 centimetres   =   175 × 10 millimetres   =   1750 millimetres

Andrea's curtain rail needs to be 1750 millimetres long.

Problems containing a mixture of units

Sometimes problems seem more complicated because they involve more than one type of unit.

Example

To make cakes for a local school event, Mary needs 1.5 kilograms (one and a half kilograms) of sugar. Each bag of sugar weighs 500 grams. How many bags does she need to buy?

We need to find out how many 500 grams are in 1.5 kilograms. The difficulty is that one of the amounts is in grams and the other is in kilograms. To make the calculation easier, we can change the 500 grams into kilograms.

1000 grams	=	1 kilogram		
500 grams	=	500 ÷ 1000	=	0.5 kilograms

So, each bag of sugar weighs 0.5 kilograms (half a kilogram). To make one and a half kilograms, Mary will need 3 bags of sugar.

 When working with measurements, always make sure that the units are the same before carrying out operations such as addition or subtraction.

Getting the relationships right

The two examples that have been used here, Andrea's DIY problem and Mary's cooking problem, illustrate one of the difficulties in converting metric units. Not only do you need to know the conversion factor, in other words how many of one unit there are in another, but you also need to know whether to multiply or divide by the conversion factor.

The diagrams below might help you:

To change:

÷ 10	÷ 100	÷ 1000
⟶	⟶	⟶

mm	cm	m	km

⟵	⟵	⟵
× 10	× 100	× 1000

Measurement

To change:

$$\div 1000 \qquad \div 1000$$

$$\longrightarrow \longrightarrow$$

mg   g   kg

$$\longleftarrow \longleftarrow$$

$$\times 1000 \qquad \times 1000$$

To change:

$$\div 10 \qquad \div 100$$

$$\longrightarrow \longrightarrow$$

mL   cL   L

$$\longleftarrow \longleftarrow$$

$$\times 10 \qquad \times 100$$

Another way of remembering whether to multiply or divide by the conversion factor is to remember that going from a smaller unit to a larger unit you divide; going from a larger unit to a smaller unit you multiply.

Practice 1

Length

1. Change 345 mm to cm
2. Change 3294 cm to m
3. Change 19 473 m to km
4. Change 90 000 cm to mm
5. Change 234 500 mm to m
6. Change 4 km to m
7. Change 2375 m to cm
8. Change 283 700 cm to km
9. Change 120 548 560 mm to km
10. Change 2.5 km to cm

Practice 2

Weight

1. Change 4560 mg to g
2. Change 20 000 g to kg
3. Change 456 g to mg
4. Change 0.5 kg to g
5. Change 450 mg to kg
6. Change 2 kg to mg

Practice 3

Capacity

1. Change 345 mL to cL
2. Change 1987 cL to L
3. Change 4.5 L to cL
4. Change 19 cL to mL
5. Change 12 345 mL to L
6. Change 0.32 L to ml

Practice 4

Fill in the blanks in the following table:

millimetres	centimetres	metres	kilometres
90 000			
	8200		
		750	
			1.5

SHAPES

Properties of shapes

In this section of the book, we are going to look at **plane shapes** – shapes that can be drawn on a flat piece of paper.

One important property of this type of shape is the number of sides it has. For example, we will see later in this section that a rectangle has four sides while a triangle has three. The lengths of the sides are also important: we will see that the width of a square is the same as its height, but that a rectangle can have a width that is different from its height.

Another important property of shapes is the **angles** that they have between their sides. An angle describes an amount of rotation. To understand this better, look at the line below:

A ——————————————— B

If this line is rotated around the point A by moving point B to a new position, we can find the amount of rotation by measuring the angle between the original and new positions of the line.

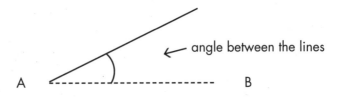

A ⟵ - - - - - - - - - - - - - - - B   ⟵ angle between the lines

Angles are measured in degrees (°) and can be measured using an angle measurer called a **protractor**.

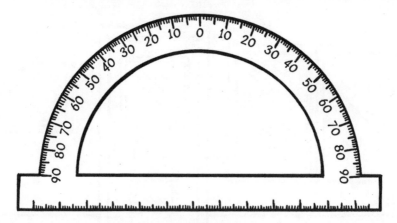

Each small division on the protractor represents 1°.

Types of angles

Right angles

Angles of 90° are often called **right angles**. Right angles are shown by drawing a small square where the two lines meet to form the angle.

When two lines meet at a right angle, one line is said to be **perpendicular** to the other.

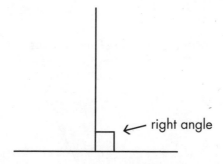

← right angle

Examples of right angles can be seen all around us.

Shapes

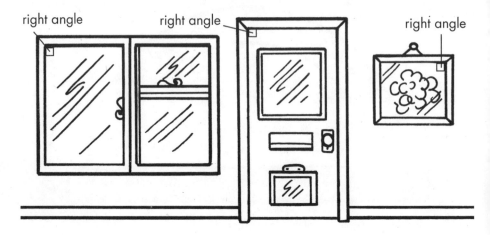

right angle   right angle   right angle

Acute angles

Angles smaller than 90° are called **acute angles**.

Obtuse angles

Angles bigger than 90° but smaller than 180° are called **obtuse angles**.

Reflex angles

Angles bigger than 180° are called **reflex angles**.

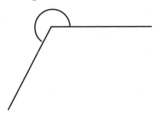

Properties of angles

In the following diagram, two lines have been drawn to meet at the point marked A.

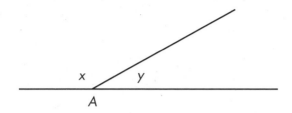

When this happens, two angles are formed, which have been marked *x* and *y* on the diagram. These two angles, whatever their respective sizes, always add up to 180°. This is an important property of angles.

 Angles on a straight line add up to 180°.

This property is true no matter how many angles are at the point. For example, if two lines meet another at the point A then three angles are formed.

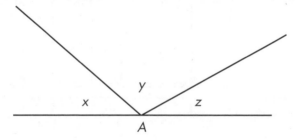

The angles *x*, *y* and *z* add up to 180°.

In the following diagram two lines have been drawn so that they **intersect** or cross each other.

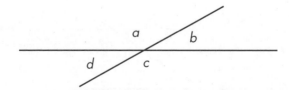

Two important properties of angles are shown here:

Opposite angles are equal: angle *a* = angle *c* and angle *b* = angle *d*.

 All four angles add up to 360°. Angle *a* + angle *b* + angle *c* + angle *d* = 360°.

Parallel lines and angles

Two or more lines are said to be **parallel** if they never intersect (cross each other), no matter how far the lines are extended. Railway tracks are a good example of parallel lines in real life.

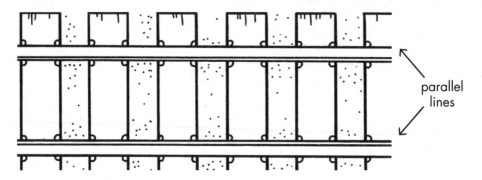

parallel lines

In the following diagram the two parallel lines are crossed by a third line.

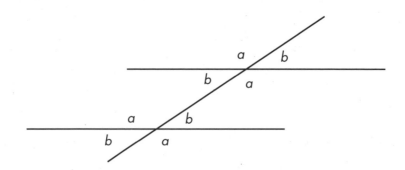

In the diagram, angles with the same letter are equal.

There are four pairs of opposite angles. Can you identify them?

The angles labelled *a* are equal and the angles labelled *b* are equal. These angles are called **vertically opposite angles**.

Also, the following angles are equal:

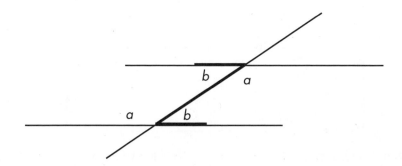

The two *b* angles are equal and the two *a* angles are equal. These angles are called **alternate angles** or **Z angles**, because the lines that form the angles make the shape of a letter Z (or a backwards Z).

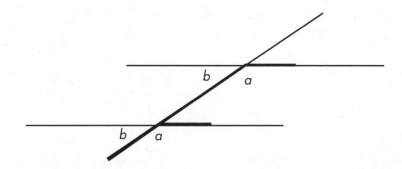

The two *b* angles are equal and the two *a* angles are equal. These angles are called **corresponding angles** or **F angles**, because the lines that form the angles make the shape of a letter F (or a backwards F).

Rectangles and squares

A **rectangle** is a four-sided figure. Its opposite sides are equal in length, and parallel. All four of its angles are right angles (90°).

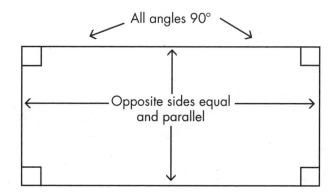

A **square** is a special type of rectangle that has all its sides equal in length.

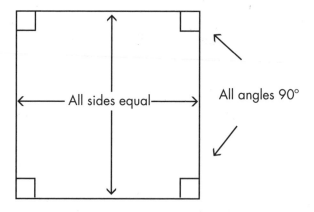

Perimeters of rectangles and squares

The **perimeter** of a plane shape is the total distance around the sides of the shapes. For a rectangle or a square, the perimeter is equal to the length of the four sides added together.

Example

A rectangular room is 3 metres wide and 7 metres long. What is the perimeter of the room?

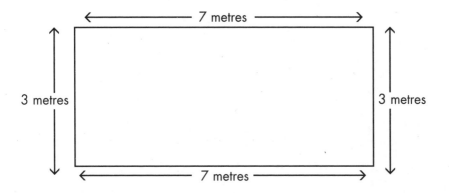

To find the perimeter we need to add the lengths of the four sides together.

Perimeter   =   3 m + 7 m + 3 m + 7 m   =   20 m

The perimeter of the room is 20 metres.

Example

Siân wants to sew a border on to a square bedspread. If each side of the bedspread is 2.5 metres long, what length of border will she need?

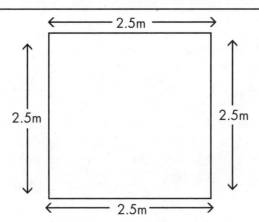

Shapes

To find the distance all the way around the bedspread we need to find the perimeter of the square.

Perimeter   =   2.5 m + 2.5 m + 2.5 m + 2.5 m   =   10 m

Siân will need 10 metres of edging for her bedspread.

 Diagrams of rectangles and squares often just show length and width on two sides, as in the diagram below. However, when you are calculating a perimeter you must remember to add all four sides, even though only two of them have been marked.

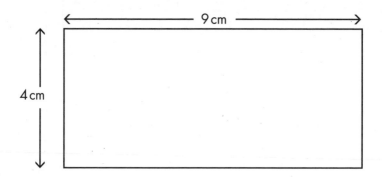

Practice 1

Find the perimeters of the following rectangles:

1. Length 10 cm, width 4 cm
2. Length 12 cm, width 6 cm
3. Length 6 m, width 2.5 m
4. Length 7 m, width 1.5 m
5. Length 60 mm, width 10 mm

Practice 2

Find the perimeters of the following squares:

1. Length of side 10 cm.
2. Length of side 5 cm.
3. Length of side 2.5 m.
4. Length of side 9 m.
5. Length of side 10 mm.

Finding perimeters with mixed units

Sometimes the length and width of a rectangle will be written with different units, as in the following example.

Example

What is the perimeter of a rectangle with length 2 metres and width 10 centimetres?

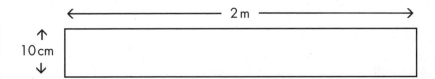

Here the length is given in metres and the width in centimetres. When adding the four sides together, we have to be careful not to confuse the units.

| Perimeter | = | 2 m + 10 cm + 2 m + 10 cm | = | 4 m 20 cm |

Or, changing the metres to centimetres before adding:

| Perimeter | = | 200 cm + 10 cm + 200 cm + 10 cm | = | 420 cm |

(1 metre = 100cm – see the section about converting metric units on page 152.)

Or, changing the centimetres to metres before adding:

| Perimeter | = | 2 m + 0.1 m + 2 m + 0.1 m | = | 4.2 m |

Shapes

What we *can't* do is add the metres and centimetres as if they were the same units: the perimeter does *not* equal 2 + 10 + 2 + 10 of any units.

Practice 3

Find the perimeters of the following rectangles:

1. Length 2 m, width 4 cm
2. Length 12 m, width 60 cm
3. Length 2.5 m, width 90 cm
4. Length 7 m, width 40 cm
5. Length 10 cm, width 60 mm

Areas of rectangles and squares

The **area** of a plane shape is a measure of the amount of space contained within the shape. In the metric system this is usually measured in square millimetres (mm^2), square centimetres (cm^2), square metres (m^2), or square kilometres (km^2).

Look at the rectangle below.

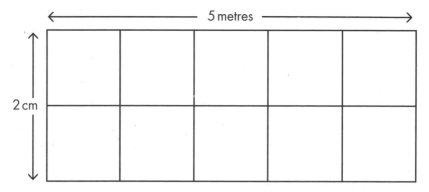

The rectangle is shown as being 5 metres long and 2 metres wide. It has been divided up into squares. Each square measures 1 metre by 1 metre, so the area of each square is a square metre ($1 m^2$).

We can calculate the area of the rectangle by adding up the squares. There are 10 squares, so the area of the rectangle is $10 m^2$ (10 square metres).

Look at the following square:

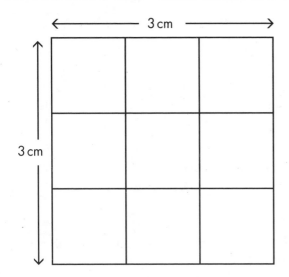

The length and width of the square are both 3 cm. It has been divided up into smaller squares. Each square measures 1 centimetre by 1 centimetre, so the area of each small square is a square centimetre (cm²).

We can calculate the area of the large square by adding up the smaller squares. There are 9 squares, so the area of the rectangle is 9 cm² (9 square centimetres).

Now think about the length and width of the rectangle. If we multiply length by width, we get the same answer for the area as we did by adding up the squares.

$$5\,m \times 2\,m = 10\,m^2$$

The same is true of the square. If we multiply the length by the width, we get the same answer for the area of the large square as we did by adding up the smaller squares.

$$3\,m \times 3\,m = 9\,m^2$$

 To find the area of a rectangle or square, multiply the length by the width.

Shapes

Practice 4

Find the areas of the following rectangles and squares. Remember to give your answers in squared units.

1. Length 20 cm, width 8 cm
2. Length 52 cm, width 6 cm
3. Length 2.5 m, width 2.5 m
4. Length 7 m, width 1.5 m
5. Length 80 mm, width 80 mm

Finding area with mixed units

If the length and width of a rectangle are given in different units, the units should be made the same before multiplying.

Example

What is the area of a rectangle with length 1.5 metres and width 20 centimetres?

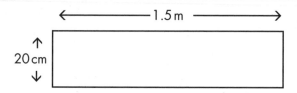

$$Area \quad = \quad 1.5\,m \times 20\,cm$$

We can't multiply together metres and centimetres, but we can change the metres to centimetres so that both measurements are in centimetres. (1 metre = 100 cm, so multiply the metres by 100 – see the section about converting metric units on page 152.)

$$Area \quad = \quad 150\,cm \times 20\,cm \quad = \quad 3000\,cm^2$$

Alternatively, we can change the centimetres to metres. (1 metre = 100 cm, so divide the centimetres by 100.)

Area = 1.5 m × 0.2 m = 0.3 m²

So the area of the rectangle is 3000 cm², or 0.3 m².

 Be careful! There are 100 cm in 1 m, but there are 10 000 cm² in 1 m².

Practice 5

Find the areas of the following rectangles. Remember to give your answers in squared units.

1. Length 2.5 m, width 8 cm
2. Length 3 m, width 50 cm
3. Length 2.5 m, width 25 cm
4. Length 7 m, width 10 cm
5. Length 10 cm, width 60 mm

Mixed problems

As we've seen before, when you are faced with a written problem it is sometimes difficult to identify what type of calculation is needed. Sometimes more than one type of calculation is required.

In the following exercise, all the questions involve finding either perimeter or area. Some of the questions ask you to use the area or perimeter value that you have found to work out something else. Read each question carefully and try to identify the key parts. You may find it helpful to use the techniques discussed in the sections on whole numbers (page 39) and fractions (page 70).

Practice 6

1. Ruth is buying carpet for her living room, which measures 4 metres by 3.5 metres. What area of carpet should she buy?

2. Terry's garden is 20 metres long and 15 metres wide. What length of fencing does he need to fence around the four sides of the garden?

3. A pack of laminated floor tiles contains 6 tiles. Each tile is 1 metre long and 60 centimetres wide. What area of floor can be covered by 1 pack?

4. A packet of grass seed contains enough seed to sow 15 square metres of lawn. How many packets will be needed to sow a lawn 9 metres long and 5 metres wide?

5. Javinda is painting a wall in her living room. The wall measures 2.5 metres by 6 metres. How many tins of paint should she buy if each tin covers 7.5 square metres?

Perimeters and areas of irregular shapes

The following examples show methods for finding perimeters and areas of irregular shapes. Notice that, although the shapes are irregular, they can all be thought of as collections of rectangles.

Example

The diagram shows the floor plan of a house. Find the total perimeter around the house.

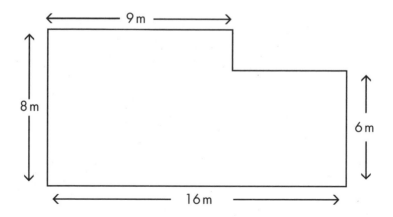

To find the perimeter of the house, we have to find the total distance all the way around the outside. We have to add up the lengths of the outside walls.

This example shows one of the problems that sometimes occurs with these kind of questions. Look at the diagram – can you see that two of the lengths are missing?

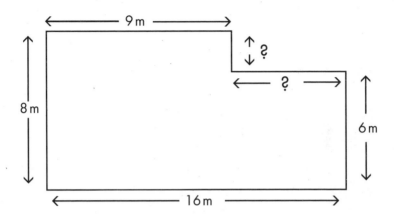

These missing lengths can be found from the other dimensions as shown below:

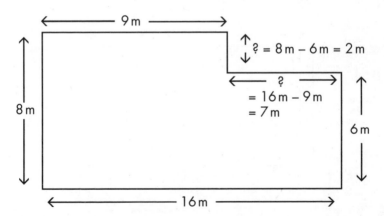

Before moving on to do the calculation, make sure you understand how the missing lengths were found.

We can now add up all the lengths to find the total perimeter around the house.

Total perimeter around the house = 8 m + 9 m + 2 m + 7 m + 6 m + 16 m
= 48 m

Example

The diagram shows the floor plan of a house. Find the total floor area of the house.

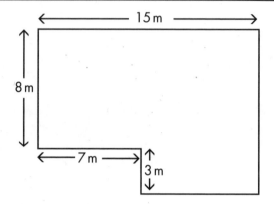

There are two ways to tackle this sort of problem.

One way is to divide the diagram into separate rectangles, find the area of each and add the areas together. There are a number of different ways of dividing up the diagram. One way is shown below:

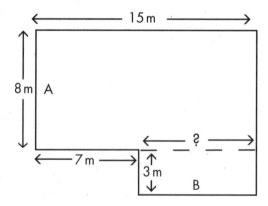

In order to find the area of rectangle A, we have to multiply its length by its width:

Area of A = 15 m × 8 m = 120 m²

In order to find the area of rectangle B, we first have to find the missing length (?).

Length of rectangle B   =   length of rectangle A – 7 m   =   15 m – 7 m
=   8 m

Look carefully at the diagram and check that you understand how and why this is being done.

Area of B   =   8 m × 3 m   =   24 m²
Total area   =   area of A + area of B   =   120 m² + 24 m²   =   144 m²

The total area of the floor plan is 144 square metres.

Alternatively, we can view the diagram as a 15 m × 11 m rectangle with a smaller 7 m × 3 m rectangle cut out.

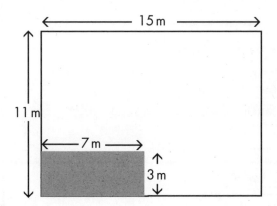

Area of large rectangle   =   15 m × 11 m   =   165 m²
Area of shaded rectangle   =   7 m × 3 m   =   21 m²
So area of floor plan   =   165 m² – 21 m²   =   144 m²

Again, we find that the area of the floor plan is 144 square metres.

Shapes

Example

The diagram shows a garden with a path, width one metre, around a rectangular lawn. Find the area of the path.

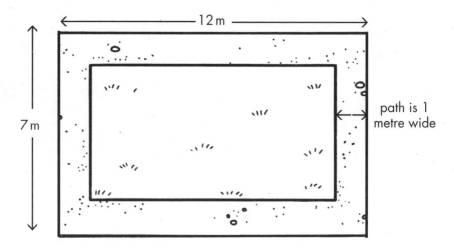

There are two ways to tackle this sort of question.

The first way is to divide the path up into four separate rectangles, find the area of each and add the four areas together.

Because we know that the path is 1 metre wide we can work out the dimensions of the four rectangles.

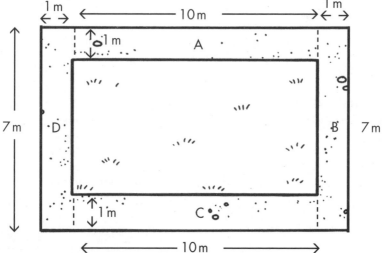

The areas of the four rectangle shown are:

Rectangle A	=	1 m × 10 m	=	10 m²
Rectangle B	=	1 m × 7 m	=	7 m²
Rectangle C	=	1 m × 10 m	=	10 m²
Rectangle D	=	1 m × 7 m	=	7 m²

Then we can work out the total area of path:

Total area $\quad = \quad$ 10 m² + 7 m² + 10 m² + 7 m² $\quad = \quad$ 34 m²

The total area of the path is 34 square metres.

Alternatively, we can find the area of the whole garden and then subtract the area of the lawn. This will leave us the area of the path.

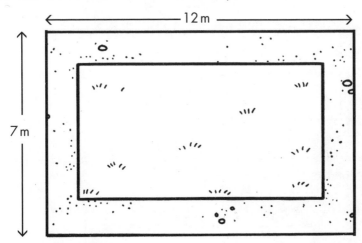

Because the path is a metre wide, the length and width of the lawn will be 2 metres shorter than the length and width of the whole garden.

Length of lawn	=	12 m – 2 m	=	10 m
Width of lawn	=	7 m – 2 m	=	5 m

Now we can calculate the areas of the lawn and the entire garden:

Area of lawn	=	10 m × 5 m	=	50 m²
Area of garden	=	12 m × 7 m	=	84 m²

Shapes

Finally, we can work out the area of the path:

Area of path = Area of garden − Area of lawn = 84 m² − 50 m² = 34 m²

Practice 7

Find the perimeters and areas of the following shapes. Remember to use square units.

1.

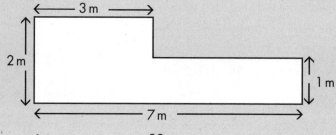

2.

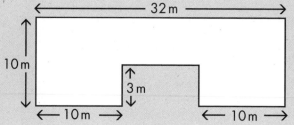

Find the areas of the shaded parts of the following shapes.

3.

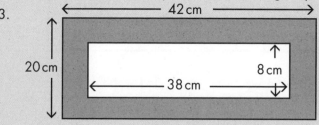

4.

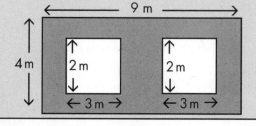

Triangles

A **triangle** is a three-sided figure. Triangles can be classified in two ways – by their sides, or by their angles.

Recognizing triangles by their sides

A **scalene triangle** has three sides of different lengths.

An **isosceles triangle** has two sides of equal length.

An **equilateral triangle** has all three sides equal.

Recognizing triangles by their angles

An **acute triangle** has all three angles less than 90°.

A **right-angled triangle** has one angle of exactly 90° (a right angle).

An **obtuse triangle** has one angle greater than 90 °.

An **equiangular triangle** has all three angles equal (this is the same as an equilateral triangle).

 The three angles of any triangle add up to 180°.

Perimeters and areas of triangles

We don't often have to calculate perimeters and areas of triangles in everyday life but it can be done.

The perimeter of a triangle can be found by adding the lengths of the three sides together.

The following example gives one practical situation in which knowing how to find the area of a triangle is useful, and shows how to calculate the area.

Example

Maura wants to paint the gable end of her house, and wants to calculate the area that she has to paint so that she knows how much paint to buy. The width of the gable end is 8 metres, the height of the walls is 10 metres and the height to the top of the roof is 15 metres. What is the area of the gable end?

The following diagram shows the dimensions of the gable end of Maura's house. The diagram shows that we can think of the gable end as a rectangle with a triangle on top. So, by calculating the areas of the rectangle and the triangle, we can work out the area of the whole gable end.

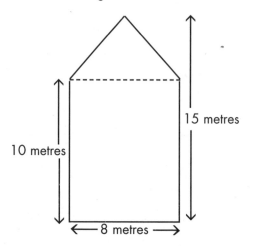

The area of a triangle is found by multiplying $\frac{1}{2}$ × base length × vertical height.

Look at the triangular part of the gable end:

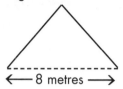

The vertical height of a triangle is the length of a line drawn from the **apex** (top) of the triangle so that it meets the base line at 90°. This is shown on the diagram below as a dotted line. In our example, the diagram of the gable end shows that the length of this line can be calculated by subtracting the height of the walls of the house from the total height of the house.

We can see from the diagram of the gable end that the total height of the house is 15 metres and that the height of the walls of the house is 10 metres. This allows us to work out the height of the triangular part of the gable:

| Height of triangle | = | 15 m − 10 m | = | 5 m |
| Area of triangle | = | $\frac{1}{2}$ × 8 m × 5 m | = | 20 m² |

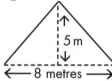

Then we can work out the area of the rectangular part of the gable wall:

| Area of rectangle | = | 10 m × 8 m | = | 80 m² |

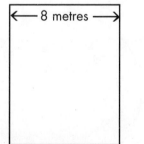

Finally, we need to add the areas of the triangle and the rectangle:

| Total area | = | 20 m² + 80 m² | = | 100 m² |

Shapes

The area of the gable end is 100 square metres.

Practice 8

Find the areas of the following triangles. Remember to use square units.

1. Base length = 8 cm, vertical height = 6 cm
2. Base length = 20 cm, vertical height = 12 cm
3. Base length = 3 m, vertical height = 1 m
4. Base length = 10 m, vertical height = 4 m
5. Base length = 100 cm, vertical height = 24 cm

Circles

A line drawn so that it is always the same distance from a fixed point produces a **circle**.

Radius

The distance from the fixed point to the line is called the **radius** (R) of the circle. It is the distance from the centre of the circle to the edge.

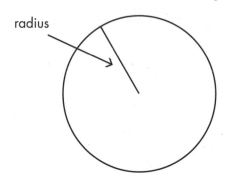

radius

Diameter

The distance across a circle through the centre is called the **diameter** (D). The diameter is twice the length of the radius.

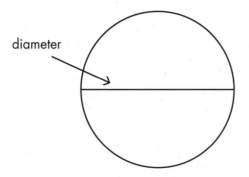

Circumference

The distance around a circle is called the **circumference** (C). The circumference of a circle is a special name for its perimeter.

Angle at the centre of a circle

Imagine a radius moving around a circle (think of the second hand of a clock). The angle through which the radius moves is the angle at the centre of the circle. It is equal to 360°.

Pi

Early mathematicians – the Babylonians, Egyptians and Greeks in particular – found that, if the circumference of any circle was divided by the diameter of that circle, the answer was always the same.

 circumference ÷ diameter = 3.1415926 ... (the number goes on for ever).

The early mathematicians couldn't measure this number to the accuracy we can today. Some of them probably wrote it as the fraction $3\frac{1}{7}$.

No matter how big or how small a circle, the circumference divided by the diameter *always* gives 3.1415926 ... This number is called **pi**, which is a letter in the Greek alphabet. The Greek letter pi is written **π**.

Finding the circumference of a circle

The circumference of a circle can be worked out using a **formula** (a special rule). If C is the circumference, π is the number mentioned above, and D the diameter, then

 $C = \pi \times D$.

Knowing where this formula comes from makes it easier to remember. We know that the circumference divided by the diameter always gives us the number π.

$C \div D = \pi$ or $C/D = \pi$

Multiplying both sides by D:

$$\frac{C}{\cancel{D}} \times \cancel{D} = \pi \times D$$

The Ds cancel out, so $C = \pi \times D$ or $C = \pi D$

Since the diameter of a circle is twice the length of the radius, then this formula can also be written as:

 $C = \pi \times 2R$ or $2 \times \pi \times R$ (often just written as $2\pi R$).

The following example shows how this might be useful in everyday life.

Example

Albert has a circular flower bed in the middle of his lawn. He wants to put decorative edging around the circumference of the bed. If the bed is 3 metres in diameter, what length of edging will he need?

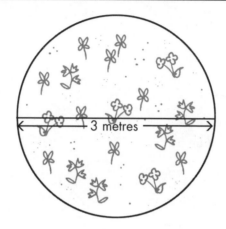

3 metres

The length of edging needed will be the same length as the circumference of the circular bed.

Circumference of bed $=$ $\pi \times 3\,m$ $=$ $3.142 \times 3\,m$ $=$ $9.426\,m$

Albert will need to buy 9.426 m of edging. In practice, he would probably have to buy 9.5 m or 10 m and cut it to size.

Note that, in this calculation, only 3 decimal places were used for the value of π. It would be pointless to use a more accurate value, since in practice nothing smaller than a millimetre could be measured.

9.426 m is equal to 9 m 42 cm 6 mm or 9 m 426 mm or 9426 mm – see the section about converting metric units on page 152.

A fourth decimal place would represent a measurement of tenths of a millimetre, which could not be measured with everyday tools.

Practice 9

Find the circumferences of the circles with the following dimensions. Use π = 3.142.

1. Diameter = 30 cm

2. Diameter = 2.5 m

3. Diameter = 300 mm

4. Radius = 10 cm

5. Radius = 90 cm

Finding the area of a circle

Imagine a circle divided up as shown:

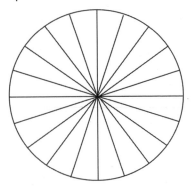

If we cut the circle in half, open up the segments and fit the two halves together, we get a shape something like this (the ends of the segments have been 'squared off'):

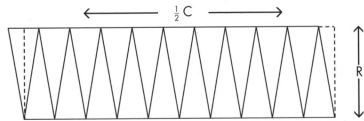

If we square off the ends of the shape (shown as dotted lines), the shape becomes a rectangle. The length of this rectangle is equal to half the circumference of the circle, because half of the segments have their base lines along the top and the other half have them along the bottom. The width of the rectangle is equal to the radius of the circle.

The area of the rectangle is a good approximation for the area of the circle. This means we can find a formula for the area of a circle by finding the area of the rectangle.

Area of rectangle   =   length × width   = $\frac{1}{2}$ C × R

We already know that the circumference of a circle is equal to π × 2R, or 2πR. Writing this instead of C we get:

Area of rectangle   = $\frac{1}{2}$ × 2πR × R   =   πR × R

R × R can be written as R^2 – the small 2 indicates how many Rs have been multiplied together.

 The area of a circle = πR^2.

It isn't necessary to remember this explanation of where the formula for the area of a circle comes from. In practice, it is enough to remember the formula and to know how to use it.

Example

Albert decides to change his circular flower bed, diameter 3 m, back into lawn. He needs to find the area of the circular bed, in order to know how much grass seed to sow.

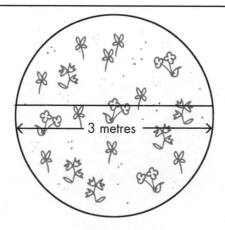

3 metres

Shapes

We know that the area of the circle = πR^2, so we need to work out the radius of the circle.

Radius of flower bed = half of diameter = 1.5 m

Putting this value into the area formula:

Area = $\pi \times 1.5\,m \times 1.5\,m$ = $3.142 \times 1.5\,m \times 1.5\,m$
     = $7.0695\,m^2$

Albert would have to buy enough seed to cover an area of $7.0695\,m^2$. In practice, he would probably have to buy seed to cover $7.5\,m^2$ or $8\,m^2$.

Practice 10

Find the areas of the circles with the following dimensions. Use $\pi = 3.142$; remember to use square units.

1. Radius = 10 cm
2. Radius = 50 cm
3. Radius = 8.5 cm
4. Diameter = 100 cm
5. Diameter = 6 m

TIME

Time is measured in **seconds**, **minutes**, **hours**, **days**, **weeks**, **months** and **years**.

60 seconds	=	1 minute
60 minutes	=	1 hour
24 hours	=	1 day
7 days	=	1 week
52 weeks	=	1 year
12 months	=	1 year

Months

The months of the year are:

January, February, March, April, May, June, July, August, September, October, November, December.

Some months have 30 days and others have 31 days. February is unusual in that it has only 28 days. Once every four years, in a **leap year**, February has 29 days.

 If you need to remember which months have 30 and 31 days, the following might help: 'Thirty days has September, April, June and November. All the rest have thirty-one, except February which has twenty-eight and twenty-nine in a leap year.'

It is very common for people to think that there are four weeks in a month. From the above, you can see that this usually isn't true (4 weeks would be 4 × 7 days = 28 days), except for February.

 Be careful when calculating weekly amounts from monthly amounts or vice versa. In calculations like this, you mustn't assume that there are four weeks in a month.

The following examples illustrate how you should tackle these calculations.

Time

Example

> Patrick earns £350 a week. How much does he earn in a month?

As discussed above, we can't multiply the weekly wage by four. Instead, we have to calculate the yearly wage by multiplying by 52 (there are 52 weeks in a year). Then we can calculate the monthly wage by dividing by twelve (there are 12 months in a year).

£350 × 52 = £18 200

Patrick's yearly or annual wage is £18 200.

£18 200 ÷ 12 = £1516.67

His monthly wage is £1516.67.

(If we had just multiplied the weekly wage by four, we would have got an answer of £1400, which is too low.)

Example

> Ron earns £2400 a month. How much does he earn in a week?

We can't divide the monthly wage by four. Instead, we have to calculate the yearly wage by multiplying by 12. Then we can calculate the weekly wage by dividing by 52.

£2400 × 12 = £28 800

Ron's yearly wage is £28 800.

£28 800 ÷ 52 = £553.85

His weekly wage is £553.85.

(If we had just divided the monthly wage by four, we would have got an answer of £600, which is too high.)

Clocks

A day is 24 hours long. The time of day is shown either on an analogue clock (a clock with hands) or on a digital clock.

12-hour (analogue) clocks

Most analogue clocks are divided into twelve large divisions, each with five smaller divisions in between. This means that there are 60 small divisions in total.

An analogue clock shows both hours and minutes on the same clock face, with one hand for each. The small hand of the clock indicates the hour, with each of the twelve large divisions showing one hour. The big hand indicates the number of minutes to or past that hour, with each of the sixty small divisions showing one minute. This means that, for the big hand, each large division shows five minutes.

The following examples show how to read different times on an analogue clock.

 12-hour times can be written with a point (dot) or a colon (:) separating the hours and minutes. For example, 1 o'clock can be written as 1.00 or 1:00. In this book, points are used throughout.

Exactly on the hour

1 o'clock (1.00)

6 o'clock (6.00)

12 o'clock (12.00)

Here the smaller hand points exactly to the 1. The larger hand points to the 12 to show that there are no minutes past or to the hour.

Here the smaller hand points exactly to the 6. The larger hand points to the 12 to show that there are no minutes past or to the hour.

Here the smaller hand points exactly to the 12. The larger hand points to the 12 to show that there are no minutes past or to the hour.

Time

Minutes past the hour

7 minutes past 3 o'clock

(3.07)

15 minutes past 8 o'clock

(8.15)

30 minutes past 10 o'clock

(10.15)

Here the smaller hand points to just past the 3. The larger hand points to 7 minutes past the hour.

Here the smaller hand points to just past the 8. The larger hand points to 15 minutes past the hour.

Here the smaller hand points to half way between the 10 and the 11. The larger hand points to 30 minutes past the hour.

Minutes to the hour

25 minutes to 4 o'clock

(3.35)

15 minutes to 8 o'clock

(7.45)

1 minute to 3 o'clock

(2.59)

Here the smaller hand points to between the 3 and the 4. The larger hand points to 35 minutes past the hour. This can be called either 35 minutes past 3 or 25 minutes to 4.

Here the smaller hand points to between the 7 and the 8. The larger hand points to 45 minutes past the hour. This can be called either 45 minutes past 7 or 15 minutes to 8.

Here the smaller hand points to just before the 3. The larger hand points to 59 minutes past the hour. This can be called either 59 minutes past 2 or 1 minute to 3.

15 minutes is equal to quarter of an hour, so 15 minutes past the hour is sometimes called 'quarter past' and 15 minutes to the hour is sometimes called 'quarter to'. 30 minutes is equal to half an hour, so 30 minutes past the hour is sometimes called 'half past'.

Morning and afternoon

With this type of clock, the day is treated as being divided into two 12-hour periods – morning and afternoon. The morning starts at 12 midnight and ends at 12 noon. The afternoon starts at 12 noon and ends at midnight.

The clock shows the time within the twelve-hour period but doesn't indicate whether it's morning or afternoon.

When we *look* at a clock we usually know which part of the day the clock is referring to. However, when we *write* times as though they were on a 12-hour clock, we must indicate whether the time refers to the morning period or the afternoon period. We do this by writing the abbreviation 'am' after the time if it is in the morning, or the abbreviation 'pm' after the time if it is in the afternoon.

 The abbreviation 'am' stands for the Latin phrase *ante meridiem*, which means 'before noon'.

 The abbreviation 'pm' stands for the Latin phrase *post meridiem*, which means 'after noon'.

Ten minutes past twelve in the afternoon would be written:

12.10 pm

Twenty-five minutes past 2 in the morning would be written:

2.25 am

Time

24-hour clocks

Most digital clocks (and a few analogue clocks) show the day divided into 24 hours. The display on the clock shows hours and minutes. The table below shows how the 24 hours of the day are displayed and their am or pm equivalents.

24-hour	12-hour
0000	12 midnight

24-hour	12-hour
0100	1.00 am
0200	2.00 am
0300	3.00 am
0400	4.00 am
0500	5.00 am
0600	6.00 am
0700	7.00 am
0800	8.00 am
0900	9.00 am
1000	10.00 am
1100	11.00 am

24-hour	12-hour
1200	12 noon

24-hour	12-hour
1300	1.00 pm
1400	2.00 pm
1500	3.00 pm
1600	4.00 pm
1700	5.00 pm
1800	6.00 pm
1900	7.00 pm
2000	8.00 pm
2100	9.00 pm
2200	10.00 pm
2300	11.00 pm

The following examples show minutes past the hour:

24-hour	12-hour	Written out in full
1435	2.35 pm	thirty-five minutes past two in the afternoon, or twenty-five minutes to three in the afternoon
0916	9.16 am	sixteen minutes past nine in the morning
1927	7.27 pm	twenty-seven minutes past seven in the evening
0004	12.04 am	four minutes past twelve in the morning

 There is no need to indicate 'am' or 'pm' when writing times using the 24-hour clock.

Changing from 12-hour to 24-hour times

From 12.00 midnight to 12.59 am: take away twelve hours, remove the point, and delete 'midnight' or 'am'.

Example

Change the following times to 24-hour times: 12.00 midnight, 12.15 am and 12.59 am.

12.00 midnight → take away 12 hours → remove point →
delete midnight → 0000

12.15 am → take away 12 hours → remove point →
delete 'am' → 0015

12.59 am → take away 12 hours → remove point →
delete 'am' → 0059

From 1.00 am to 12.59 pm: remove the point, and add zero at the beginning if the hours are in single figures.

Example

Change the following times to 24-hour times: 1.00 am, 3.30 am, 12.00 pm and 12.59 pm.

1.00 am → remove point → add zero at beginning → 0100

3.30 am → remove point → add zero at beginning → 0330

12.00 pm → remove point → 1200

12.59 pm → remove point → 1259

From 1.00 pm to 11.59 pm: add 12 hours and remove the point.

Example

Change the following times to 24-hour times: 1.00 pm, 3.46 pm, 7.28 pm and 11.59 pm.

1.00 pm	→	add 12 hours	→	remove point	→	1300
3.46 pm	→	add 12 hours	→	remove point	→	1546
7.28 pm	→	add 12 hours	→	remove point	→	1928
11.59 pm	→	add 12 hours	→	remove point	→	2359

Changing from 24-hour to 12-hour times

From 0000 to 0059: add 12 hours, insert a point between the hours and the minutes, and write 'am' or 'midnight'.

Example

Change the following times to 12-hour times: 0000, 0015 and 0059.

0000 → add 12 hours → insert point → write 'midnight' →
12.00 midnight

0015 → add 12 hours → insert point → write 'am' →
12.15 am

0059 → add 12 hours → insert point → write 'am' →
12.59 am

From 0100 to 1159: insert a point between the hours and the minutes, remove the zero from the beginning, and write 'am'.

Example

Change the following times to 12-hour times: 0100, 0954 and 1030.

0100	→	insert point	→	remove zero	→	write 'am'	→	1.00 am
0954	→	insert point	→	remove zero	→	write 'am'	→	9.54 am
1030	→	insert point	→	remove zero	→	write 'am'	→	10.30 am

From 1200 to 1259: insert a point between the hours and the minutes, and write 'pm'.

Example

Change the following times to 12-hour times: 1200, 1220 and 1259.

1200	→	insert point	→	write 'pm'	→	12.00 pm
1220	→	insert point	→	write 'pm'	→	12.20 pm
1259	→	insert point	→	write 'pm'	→	12.59 pm

From 1300 to 2359: take away 12 hours, insert a point between the hours and minutes, and write 'pm'.

Example

Change the following times to 12-hour times: 1300, 1642 and 2359.

1300 → take away 12 hours → insert point → write 'pm' → 1.00 pm

1642 → take away 12 hours → insert point → write 'pm' → 4.42 pm

2359 → take away 12 hours → insert point → write 'pm' → 11.59 pm

Practice 1

Change the following times into times in the 24-hour clock:

1. 3.20 pm
2. 7.05 am
3. 6.35 pm
4. 7.59 am
5. 10 am

Practice 2

Change the following times into times in the 12-hour clock:

1. 1245
2. 0321
3. 1723
4. 1400
5. 0046

Time

Calculating time passed

It is often useful to be able to calculate the difference between two times.

> **Example**
>
> Ralph's train was due at 1345 but is now expected at 1417. How long will he have to wait?

Here we have to find the difference between the two times, or how much time has passed from the first time to the second.

We could try to do this calculation by taking one time from the other in the usual way – by putting one number under the other and subtracting. However, a difficulty occurs if we have to borrow from one column to another.

To illustrate this difficulty we will try and subtract in the usual way.

We want to take 1345 from 1417. Remember that the first two figures of the numbers represent hours and the third and fourth numbers represent minutes

$$
\begin{array}{r}
\text{hours} \quad \text{minutes} \\
1\,4 \quad\ \ 1\,7 \\
-\ \ 1\,3 \quad\ \ 4\,5 \\
\end{array}
$$

When we try to subtract the minutes (17 – 45), we can't because 45 is bigger than 17. To make it possible to subtract, we have to 'borrow' from the hours. (See the section about subtracting with borrowing on page 15 in the section about whole numbers.) But because there are 60 minutes in an hour, the one hour we are borrowing becomes 60 minutes. We have to add 60 minutes on to the 17 minutes already there.

$$
\begin{array}{r}
\overset{\overset{\displaystyle 60}{\to}}{1\,4}\,1\,7 \\
-\ \ 1\,3\,4\,5 \\
\hline
2 \\
\end{array}
$$

The 17 minutes become 77 minutes when the 60 is added on. The 4 hours reduces to 3 hours because we have 'borrowed' one. Then we can subtract:

$$
\begin{array}{r}
1\ 3\ 7\ 7 \\
-\ \underline{1\ 3\ 4\ 5} \\
3\ 2
\end{array}
$$

Ralph would have to wait 32 minutes for his train.

This method often causes difficulty. Normally, when we subtract, the number borrowed is a ten, a hundred, or a thousand. With time, however, we have to borrow sixty and that can be confusing.

A better method for this type of calculation is shown below.

We want to find out how much time will elapse between 1345 and 1417.

1345   →   ?   →   1417

It is easier if we do this in two steps.

Step 1: calculate how many minutes up to the next full hour.

1345   →   15 minutes   →   1400

Step 2: calculate how many minutes from 1400 up to the time required.

1400   →   17 minutes   →   1417

The total time elapsed will be the times found in the two steps added together.

15 minutes + 17 minutes = 32 minutes

Sometimes three steps are required as shown in the next example.

Example

Marjorie leaves home at 1.45 pm and cycles to a friend's house. She arrives at 4.16 pm. How long did her journey take?

Step 1: calculate how many minutes up to the next full hour.

1.45 pm   →   15 minutes   →   2.00 pm

Step 2: calculate how many hours up to the hour of her arrival time.

2.00 pm   →   2 hours   →   4.00 pm

Step 3: calculate how many minutes from 4.00 pm up to the arrival time.

4.00 pm   →   16 minutes   →   4.16 pm

The total time elapsed will be the times found in the three steps added together.

15 minutes + 2 hours + 16 minutes = 2 hours 31 minutes

So Marjorie's journey took 2 hours and 31 minutes.

You will see from the two examples that the step method can be used with 12-hour or 24-hour times.

Practice 3

Find how long it is between the following times. All are on the same day.

1. 2.36 pm and 6.00 pm
2. 6.56 am and 9.00 am
3. 11.15 am and 3.45 pm
4. 12 midnight and 3.38 am
5. 1.05 am and 11.55 am

6. 1245 and 1800
7. 0734 and 1300
8. 1906 and 2355
9. 0054 and 1621
10. 1703 and 2230

TEMPERATURE

Measuring temperature

We can measure the length of something using a ruler or a tape measure, and we can measure the weight of something using a weighing scale. In the same way, we can measure temperature – the measurement of how hot something is – using a **thermometer**.

The most common type of thermometer consists of a glass tube containing a coloured liquid – usually alcohol or mercury – that expands and rises up the tube as the surrounding temperature increases. A scale, marked either on the glass itself or alongside it, allows the temperature to be read off from the top of the liquid. (See the section about scales on page 209 in the data section of this book.)

 The divisions on the scale are called degrees. The ° symbol is used to show degrees.

Medical staff now often use digital thermometers, which work by converting the temperature (of a patient's body, for example) into an electrical current. The temperature is then shown as a digital read-out.

Most thermometers are **calibrated** (marked off) using one of the two temperature scales described below.

The Fahrenheit scale

This was invented by the German physicist Gabriel Fahrenheit (1686–1736). There are a number of versions of the story of how he calibrated his scale. The most widely accepted is that he chose as a starting point for his scale the coldest recorded temperature measured in the winter of 1708 in his home town of Gdansk (now in Poland). He made this the zero mark (0°F) on his scale. (The modern day equivalent of this in Celsius would be −17.8°C.) He later reproduced this result using a mixture of ice, ammonium chloride and water. His own body temperature he fixed (slightly inaccurately) as 100°F.

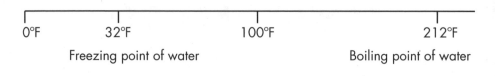

0°F   32°F   100°F   212°F

Freezing point of water   Boiling point of water

Using these as fixed points on his scale made the freezing and boiling points of water 32°F and 212°F respectively. Between the freezing point and the boiling point of water there are 180 degrees.

Celsius scale

This was invented by the Swedish astronomer Anders Celsius (1701–44). He used the freezing and boiling points of water as the start and end points of his scale. He divided his scale into 100 divisions or degrees. On the Celsius scale, water freezes at 0 degrees Celsius (0°C) and boils at 100 degrees Celsius (100°C).

The Celsius scale is part of the metric system of measurement. At one time it was called the centigrade scale because it has 100 divisions (*centi-* means 'a hundredth' – see page 151 in the section about metric measurement.)

The Celsius scale has been adopted by most countries in the world. The USA is a notable exception; there the Fahrenheit scale is still widely used. In the UK the Celsius scale has gradually replaced the Fahrenheit scale, although both are sometimes used. For example, weather reports often give temperatures in both scales.

Converting from Fahrenheit to Celsius

To change a temperature in degrees Fahrenheit (°F) to degrees Celsius (°C):

 Subtract 32

 Multiply the answer by 5

 Divide the answer by 9

Example

Change 86°F to degrees Celsius (°C).

Following the above steps:

    86 – 32 = 54
    54 × 5 = 270
    270 ÷ 9 = 30

So 86°F is 30°C.

Converting from Celsius to Fahrenheit

To change a temperature in degrees Celsius (°C) to degrees Fahrenheit (°F):

 Multiply by 9

 Divide the answer by 5

 Add 32 to the answer

Example

Change 25°C to degrees Fahrenheit (°F).

Following the above steps:

    25 × 9 = 225
    225 ÷ 5 = 45
    45 + 32 = 77

So 25°C is 77°F.

When converting temperatures, the answer is rarely a whole number as in the above examples. If the answer is not a whole number then round it off to the nearest whole number. (See page 33 in the section about whole numbers if you aren't sure.)

Temperature

Practice 1

Change the following temperatures to degrees Celsius:

1. 59°F
2. 68°F
3. 77°F
4. 149°F
5. 32°F

Practice 2

Change the following temperatures to degrees Fahrenheit:

1. 10°C
2. 35°C
3. 30°C
4. 15°C
5. 100°C

Temperatures below freezing

On the Fahrenheit scale, temperatures below freezing are shown as numbers less than 32°F. For example, 22°F is 10 degrees below freezing.

On the Celsius scale, temperatures below freezing are shown as negative numbers. For example, –5°C is 5 degrees below freezing. (See page 35 in the section about whole numbers if you are confused about negative numbers.)

DATA

The importance of data

The collection, display and analysis of **data** or information is becoming an increasingly important part of our daily lives. Information is constantly being collected about our lives and the world around us and displayed in tables, charts and graphs. We are at a disadvantage if we do not understand a little about how this information is collected and, more importantly, what it means when it is displayed in these different ways.

The collection, display, and analysis of data is a vast field, far too big to be covered in depth in this book. In this section we will look at the basic principles behind the collection of data, some of the more common methods of displaying data, and some of the methods used to analyse data.

Types of data

There are two main types of data: **quantitative** and **qualitative**.

Quantitative data consists of information that can be recorded as numbers. Examples include the heights of children in a class, the number of people living in the UK, the number of cars passing a busy road junction, or the number of patients visiting a doctor's surgery.

Qualitative data consists of information that can be recorded as words, such as opinions, voting intentions, or favourite television programmes.

Collecting data

Data can be collected in many ways. Some examples of collecting methods are:

 Direct observation

 Surveys

 Questionnaires

CCTV cameras and other technological methods

When information is first collected it is called **raw data**. It can look like no more than a jumble of numbers and can be difficult to understand. There are various ways we can display data to make the information easier to understand.

Stem-and-leaf diagram

A **stem-and-leaf diagram** is a useful way of showing a lot of numerical data in one simple diagram.

The following stem-and-leaf diagram shows the results from a survey of patients' ages in a doctor's surgery, taken over a week.

```
0 | 4 9 2 8 3 9 8 6 4 8 6
1 | 9 3 7 2 8 3 0 8 6 7 3 4 2 1
2 | 0 5 9 4 3 8 2 7 8 7 3 9 4 7 2
3 | 3 0 2 1 8 4 9 7 5 6 1 2 3 4 2 1 2 4
4 | 3 5 7 9 9 0 6 8 5 4 5 9 8
5 | 9 8 5 4 2 5 8 2 6 7 8 0 3 5 4 3 6 4 9 7 2 1
6 | 9 1 6 7 0 8 4 7 2 8 5 0 7 8
7 | 7 2 6 5 7 8 5 6 3 4 2
8 | 3 7 8 3 0 9 8 8 9 3 2 1
9 | 4 3 8 7 5 4
```

In a stem-and-leaf diagram, the numbers to the left of the line (the **stem**) are common to all the numbers to the right of the line (the **leaves**) in that particular row.

The numbers in the row beginning with 0 represent the ages of all the patients under 10. We can see that there was a 4 year old, a 9 year old, a 2 year old and so on, working across the line of numbers in the first row from left to right.

The numbers in the row beginning with 1 represent the ages of all the patients who are 10 or over but who are under 20. We can see that there was a 19 year old, a 13 year old, a 17 year old and so on, working across the line of numbers in the second row from left to right. Notice how the 1 is used as a common 'ten' for all the numbers in that row.

The numbers in the row beginning with 2 represent the ages of all the patients who are 20 or over but who are under 30. We can see that there was a 20 year old, a 25 year old, a 29 year old and so on, working across the line of numbers in the third row from left to right. Notice how the 2 is used as a common 'twenty' for all the numbers in that row.

… and so on down to the last row, which represents the ages of the patients who are 90 or over.

This stem-and-leaf diagram makes it easy to see how many people are in each age group. For example, counting up the numbers to the right of the line in the row beginning with a 3 tells us that 18 people in their thirties visited the doctor that week.

It also makes it easy to see how many people there are below or above a certain age. For example, counting up the numbers to the right of the line in all the rows up to the one beginning with 6 tells us that 93 people under sixty visited the doctor. Counting up the numbers to the right of the line in all the rows from the one beginning with 6 to the last row tells us that 43 people who are sixty or over visited the doctor.

Constructing a stem-and-leaf diagram

The following example shows how to construct a stem-and-leaf diagram.

Example

A survey of the ages of people using a public library in one day produced the results shown below.

19 56 21 33 10 24 59 46 29 37 15 43 9 18 24 11 59 37
50 40 14 82 14 6 5 20 45 21 25 56 37 16 23 18 29 38
46 39 16 54 36 51 24 43 8 41 24 35 32 41 52 53 13 15
17 32 42 19 37 50 14 17 28 42 15 30 60 7

Construct a stem-and-leaf diagram of the data.

The stem-and-leaf diagram sorts the numbers into ages below ten, ages from ten to nineteen, ages from twenty to twenty-nine, and so on.

Writing down the ages of people below ten: 9 6 5 8 7

Writing down the tens: 19 10 15 18 11 14 14 16 18 16 13 15 17 19 14 17 15

Writing down the twenties: 21 24 29 24 20 21 25 23 29 24 24 28

Writing down the thirties: 33 37 37 37 38 39 36 35 32 32 37 30

Writing down the forties: 46 43 40 45 46 43 41 41 42 42

Writing down the fifties: 56 59 59 50 56 54 51 52 53 50

Writing down the sixties: 60

Writing down the seventies: (none)

Writing down the eighties: 82

 When doing this, it is very easy to miss numbers. It helps to cross numbers off once they have been recorded.

The ages are now put into a stem-and-leaf diagram.

```
0 | 9  6  5  8  7
1 | 9  0  5  8  1  4  4  6  8  6  3  5  7  9  4  7  5
2 | 1  4  9  4  0  1  5  3  9  4  4  8
3 | 3  7  7  7  8  9  6  5  2  2  7  0
4 | 6  3  0  5  6  3  1  1  2  2
5 | 6  9  9  0  6  4  1  2  3  0
6 | 0
7 |
8 | 2
```

As before, the diagram allows us to analyse the data more easily.

For example, if we wanted to know how many people under twenty visited the library on that day, we simply have to count up the figures to the right of the line in the first two rows.

If we wanted to know how many people aged 60 or over visited the library, we count up the number of figures to the right of the line in the last three rows.

Tables

A table is another useful way of showing a lot of numerical data in one simple diagram.

Look at this record of the number of different birds seen in someone's garden each month. The numbers of blackbirds, thrushes, sparrows, and robins have been recorded:

January: 56 blackbirds, 34 thrushes, 22 sparrows, 15 robins; February: 55 blackbirds, 32 thrushes, 12 sparrows, 18 robins; March: 70 blackbirds, 45 thrushes, 48 sparrows, 20 robins; April: 70 blackbirds, 65 thrushes, 32 sparrows, 19 robins ... and so on for the remaining twelve months of the year.

Looking at these results, we can see that it is going to be difficult to identify any patterns, or to calculate totals of different birds at the end of the year. The way the information has been written doesn't make it easy to analyse. Instead, we can reorganize the information into a simple two-way table as shown below.

	Jan	Feb	Mar	April	May	June	July	Aug	Sept	Oct	Nov	Dec
Blackbirds	56	55	70	70	90	123	180	98	65	54	45	30
Thrushes	34	32	45	65	86	99	166	84	70	46	39	27
Sparrows	22	12	48	32	75	84	91	65	43	21	15	10
Robins	15	18	20	19	45	67	34	54	39	32	31	28

Arranged like this it is easier to find, for example, the total number of birds in each month – add the numbers in the columns. For example, we can work out the number of birds recorded in May:

Total birds in May $= 90 + 86 + 75 + 45 = 296$

Or we can work out the total number of birds recorded in December:

Total birds in December $= 30 + 27 + 10 + 28 = 95$

It is also easier to find the total number of any one kind of bird over the whole year – add the numbers in the rows. For example, we can work out the total number of blackbirds recorded:

Blackbirds recorded $= 56 + 55 + 70 + 70 + 90 + 123 + 180 + 98 + 65 + 54 + 45 + 30$

$= 936$ blackbirds recorded

Robins recorded $= 15 + 18 + 20 + 19 + 45 + 67 + 34 + 54 + 39 + 32 + 31 + 28$

$= 402$ robins recorded

With information arranged in a table like this, it is often easier to see patterns and trends.

Scales

Some ways of displaying data use **scales**. So, before looking at these ways of displaying data, it will be useful to understand how a scale is constructed and read.

 A scale is a system of ordered marks at fixed intervals, used as a reference standard in measurement.

Data

The picture shows some examples of scales that we might see in everyday life. The scales on these instruments allow us to measure length and temperature.

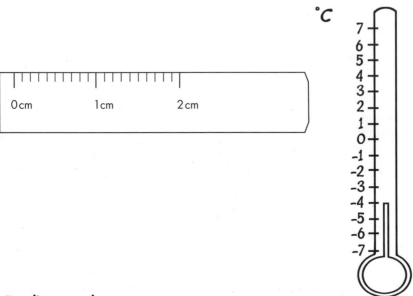

Reading a scale

In the following diagram, the line has been divided up at regular intervals to create a scale. The starting and end points of the scale are known (10 and 20), but the points in between are not.

Most scales are like this: some of the divisions are known and others are not. To read the scale, we need to be able to work out the value of the unknown divisions.

Example

In the following diagram the arrow is pointing at one of the unknown divisions. What is its value?

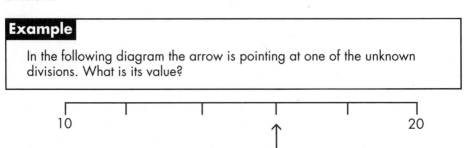

Sometimes, is it is possible to look at a scale and 'see', without calculating, what the unknown divisions represent. So, perhaps you were able to look at the diagram and 'see' that the arrow is pointing at a value of 16.

However, don't worry if you couldn't: the following method will allow you to work out the missing divisions on any scale.

Step 1: choose two numbers that are next to each other on the scale and that are known. In the above diagram, the values of 10 and 20 are known and next to each other (ignoring the unknown divisions in between).

Step 2: take the smaller number from the bigger number.

$20 - 10 = 10$

This gives a value for the 'distance' between them.

Step 3: count the number of *spaces* between them. Not the marks, but the spaces.

There are 5 spaces

Step 4: divide the difference between the two numbers found in step 2 by the number of spaces.

$10 \div 5 = 2$

This gives us a value for each space.

We say the scale is 'going up in twos'. Or, each division represents an **increment** (an increase) of 2.

Once this increment is known, we can work out the value of each of the unknown divisions.

Step 5: starting at the lower of the known numbers, count up in twos.

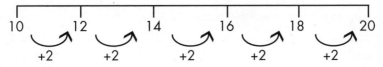

Bar charts

A **bar chart** is another useful way of showing data. Have a look at the bar chart below. It shows the results of a traffic survey carried out by a group of parents who are campaigning to get a zebra crossing outside their local school. The parents counted the number of cars passing the school each hour from 8am until 5pm.

Bar chart showing traffic survey

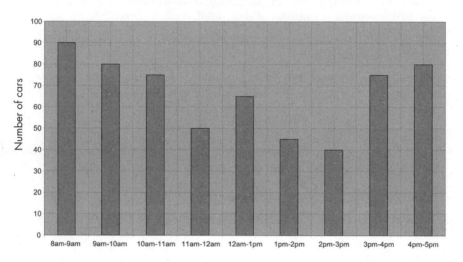

Reading a bar chart

In the bar chart above, we can 'read' how many cars were recorded in a particular hour.

Example

> Use the bar chart to find out how many cars were recorded between 10 am and 11 am.

To 'read' how many cars were recorded between 10 am and 11 am, find the bar for that hour and look across from the top of the bar to read the number of cars.

75 cars were recorded between 10 am and 11 am

Constructing a bar chart

The following example shows how to construct a bar chart.

Example

> In a well-known quiz show the audience were asked to help the contestant by voting for answers A, B, C or D. The percentages voting for each answer are shown in the table. Construct a bar chart of the data.

A	B	C	D
40	35	5	20

A bar chart is drawn on grid or graph paper. The steps involved are shown below.

Step 1: draw a horizontal axis and a vertical axis on which the two rows of data shown in the table can be represented.

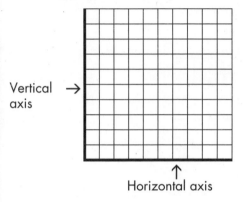

Vertical → axis

Horizontal axis

Step 2: mark off both axes so that each one represents one of the rows of data in the table.

Usually the data in the top row of the table goes along the horizontal axis and the data in the bottom row of the table goes up the vertical axis. In a bar chart the the horizontal axis is usually marked off with **categories** of some kind. You can write these underneath the horizontal axis. You should write them so that there are equal spaces between them.

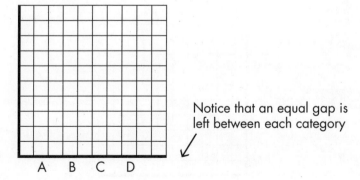

Notice that an equal gap is left between each category

A   B   C   D

The values in the second row of the table are written on the vertical axis. A common mistake is to space out the actual numbers that appear in the table in the same way as we did with the categories on the horizontal axis. This is incorrect, because the numbers in the bottom row of the table are not categories. They are **values** that are attached to the categories A, B, C and D. What we have to do is scale the axis so that all the numbers in the table are represented. (See the section above on scales.)

 It is usually best to start the scale of an axis at zero. The zero is usually written at the junction of the two axes.

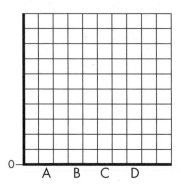

The largest percentage value in the table is 40, so the scale on the vertical axis has to go up to at least 40. The vertical axis has to be divided up into equal intervals between 0 and 40. There are a number of ways of doing this, and the person drawing the graph has to decide which is most suitable.

If each square represents an increment of 5 then the vertical scale would look like this:

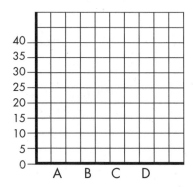

Step 3: draw bars or blocks to represent the value of each category. Make sure that the bars or blocks are spaced evenly apart.

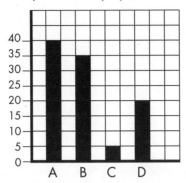

Step 4: label the axes to explain what the numbers and the categories A, B, C and D stand for. Otherwise the bar chart is meaningless! You should also give the bar chart a heading to explain what it is showing. The completed bar chart would look like this:

Bar chart showing percentage voting

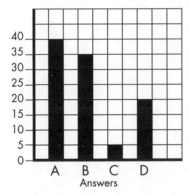

Line graphs

Some information is better illustrated by using a **line graph**. Line graphs represent data by points on a grid rather than by using blocks. The points are joined together with a continuous line. This can be a straight line or a curve. The line graphs we will be looking at are straight line graphs. Line graphs are useful for showing trends or changes in data.

Data

Constructing a line graph

The following example shows how to construct a line graph.

> ### Example
>
> Information has been collected over a ten-year period about the change in the heights of 10-year-old boys living in an inner-city area of London. The table below shows the data (rounded to the nearest centimetre). Draw a line graph to represent this information.

Year	1996	1997	1998	1999	2000	2001	2002	2003	2004	2005
Height (cm)	180	182	175	185	185	179	189	190	200	200

To draw a line graph to illustrate this information, we follow steps 1 and 2 shown above for drawing a bar graph.

Step 1: draw axes on graph paper.

Step 2: add scales to the axes.

Notice that the scale on the vertical axes starts at 175cm. In this example, starting with the smallest height rather than zero will not affect the accuracy or usefulness of the graph and will make it easier to construct a scale on the vertical axis.

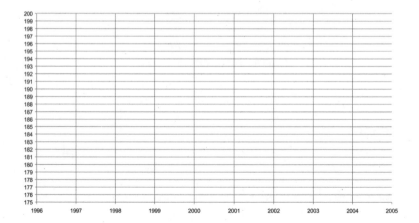

Step 3: instead of drawing bars as in the bar chart, plot points on to the grid corresponding to the height for each year. For example, above 1996 and opposite 180 cm, mark in a point or a *very small* cross. Do the same above 1997 and opposite 182 cm. Repeat for the remaining years and heights. Then use a ruler to join the points with straight lines. Remember to add labels to the axes and give the graph a title.

Heights of 10-year-old boys

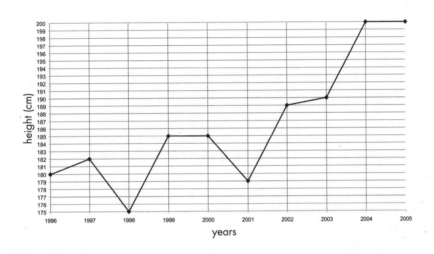

years

Using line graphs

A line graph like this allows us to see what is happening to data over a period of time.

In this example, we can see that the heights of the children rise and fall over the ten years, but the overall **trend** is upwards. The average height of this age group seems to be increasing.

The graph can also be used to get information about the heights of the children which is not contained in the original table.

Example

In which year did the average height reach 188cm?

We can find 188cm on the vertical scale and draw a line across until we reach the graph. Drawing a line down from this point to the time axis will tell us the year in which the average height reached 188cm.

Heights of 10-year-old boys

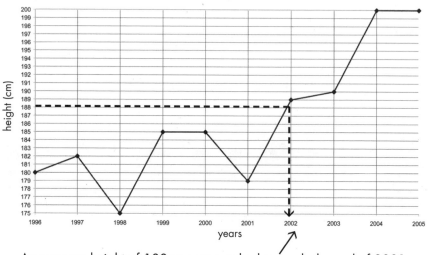

An average height of 188cm was reached towards the end of 2001

This method of 'tracking' from the vertical axis to the graph and down to the horizontal axis, or up from the horizontal axis to the graph and across to the vertical axis, is a very useful technique for obtaining information from a straight line graph, as shown by the next example.

The following straight line graph is a **conversion graph** that can be used to change from pounds (£) to dollars ($).

Conversion graph £ to $

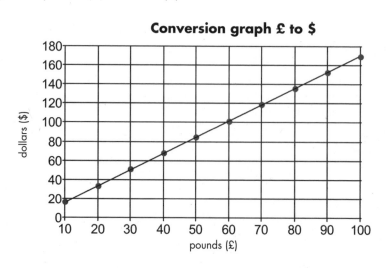

The conversion graph can be used to change pounds into dollars or dollars into pounds.

Change $40 into pounds using the conversion graph.

To change $40 into pounds, find 40 on the dollar scale, track across to the graph, and then track vertically downwards to the pound axis, as shown below.

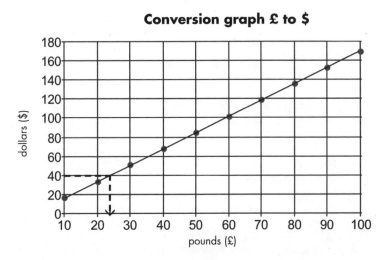

Conversion graph £ to $

The arrow shows that $40 is approximately £23. If the graph was drawn on graph paper then we could read this value more accurately.

Change £65 into dollars using the conversion graph.

To change £65 into dollars, find 65 on the pound scale, track vertically upward to the graph, and then track across to the dollar axis, as shown below.

Conversion graph £ to $

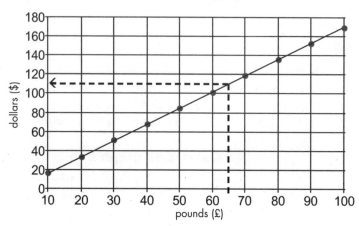

The arrow shows that £65 is approximately $110. If the graph was drawn on graph paper then, again, we could read this value more accurately.

 In some cases, plotting data as a series of points gives a curve instead of a straight line.

Pie charts

Have a look at the table showing the Smith family's weekly spending:

Food	Rent	Bills	Travel	Other
£100	£80	£50	£75	£55

We can use a **pie chart** to illustrate their spending. The size of each 'slice' of the pie chart indicates how much has been spent on that item.

The pie chart shows that most money was spent on food. The smallest amount of money was spent on bills.

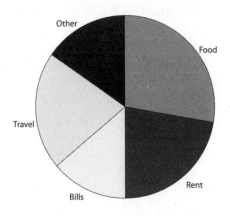

Other

Food

Travel

Rent

Bills

Drawing a pie chart

The next example shows how to draw a pie chart.

Example

In a survey, people were asked to name their favourite soap opera. The following table shows the results of the survey. Draw a pie chart of the results.

Neighbours	EastEnders	Coronation Street	Hollyoaks
35	55	50	40

Step 1: draw a circle using a pair of compasses, and mark the centre.

The size of the circle is not important. Common sense and the size of your paper will tell you how large to draw it. Don't make it too small otherwise it becomes difficult to read.

Step 2: calculate the size of the slices so that they accurately represent the number of people in each category.

 The size of each slice will depend on the angle that the slice makes at the centre of the circle.

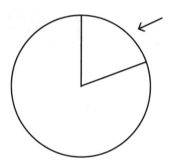

The size of the slice depends on the size of this angle

The total angle at the centre of any circle is 360° (see the section about circles on page 182).

The whole of the pie represents the total number of people in the survey, so the total number of people in the survey is represented by the angle 360°.

We need to find the total number of people in the survey. We can do this by adding together the numbers of people in each category.

Total number of people = 35 + 55 + 50 + 40 = 180

We now know that 180 people are represented by 360°. We can use this information to work out the angle that represents one person, by dividing 360° by the number of people.

Angle representing 1 person = 360° ÷ 180 = 2°

 This is the key calculation when constructing a pie chart. Make sure you fully understand it before going on.

Once the angle for one person has been found, the angles for the number of people in each category can be found by multiplication.

Neighbours:	2° × 35 = 70°
EastEnders:	2° × 55 = 110°
Coronation Street:	2° × 50 = 100°
Hollyoaks:	2° × 40 = 80°

The sum of the all the angles should add up to 360°. This is a useful check to make sure you have calculated the angles correctly. If the total isn't 360° then you should check your calculations again.

Step 3: draw the slices of the pie with the correct angles.

First draw a line from the edge of the circle to the centre as shown. This gives a line from which you can start measuring the angles.

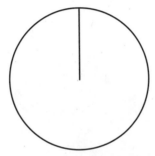

Using a protractor, carefully measure the first angle and draw a line to the centre to make the first slice of the pie.

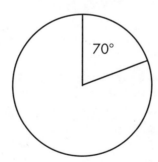

Move the protractor around and measure the angle for the next slice of the pie.

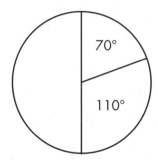

Continue around the circle, until all slices have been drawn. You should find that the final slice fits exactly into the space remaining. If it doesn't, go back and check your angle measurements.

The finished pie chart will look like this:

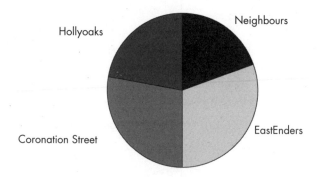

If a pie chart is to go in a document, perhaps for a project or presentation, you would normally leave out the angles.

The slices of the pie chart can be coloured or shaded. The label for each category can be written alongside or inside the slice. Alternatively, a **key** can be added to explain what each colour or shading represents.

Key

■ Neighbours
□ EastEnders
■ Coronation St
■ Hollyoaks

Rounding off angles

Sometimes, when carrying out these calculations, decimal answers may be obtained, as shown in the following example.

> ### Example
>
> The table shows the average daily hours of sunshine in some Scottish cities in August 2006. Use the data to draw a pie chart.
>
> | Edinburgh | Glasgow | Aberdeen | Inverness |
> |:---------:|:-------:|:--------:|:---------:|
> | 8 | 7.5 | 9.25 | 7 |

Calculate the angles for the pie chart:

Total hours = 8 + 7.5 + 9.25 + 7 = 31.75

Work out the angle that represents one hour:

Angle for 1 hour = 360° ÷ 31.75 = 11.34° (to 2 decimal places)

Now work out the angle for each slice of the pie:

8 hours:	11.34° × 8	=	90.72°
7.5 hours:	11.34° × 7.5	=	85.05°
9.25 hours:	11.34° × 9.25	=	104.9°
7 hours:	11.34° × 7	=	79.38°

Using an ordinary angle measurer (protractor) it is impossible to draw these angles to these degrees of decimal accuracy. Instead, in this situation we round off the angles to the nearest degree. (See the section about rounding off on page 33.)

8 hours:	11.34° × 8	=	90.72° which rounds off to 91°
7.5 hours:	11.34° × 7.5	=	85.05° which rounds off to 85°
9.25 hours:	11.34° × 9.25	=	104.9° which rounds off to 105°
7 hours:	11.34° × 7	=	79.38° which rounds off to 79°

The rounded-off angles are the angles that are drawn.

If you are checking your calculations by adding the rounded-off angles, then the total might be *slightly* more or less than 360°. A difference of about 1 or 2 degrees can be ignored, as it can be put down to the rounding off rather than to any miscalculation.

In this example, the rounded-off angles add up to exactly 360°:

91° + 85° + 105° + 79° = 360°

Reading a pie chart

The following two examples show how information can be found from a pie chart.

Example

The pie chart below shows the results from a survey of people in England and Wales who were asked how they would vote in the next general election. If 180 people took part in the survey, and the angle for the Liberal Democrat 'slice' is as shown, find how many people said they would vote for the Liberal Democrats.

England and Wales

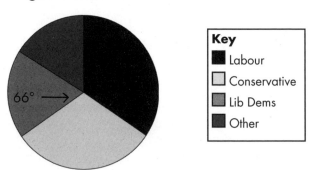

First, we need to find the number of people represented by 1° of the pie chart.

360°   =   180 people

So 1°   =   180 ÷ 360 =   0.5 people (don't worry about the absurdity of this answer!)

The angle for the Liberal Democrat slice is 66°, so we can now work out the number of people represented by that angle.

Number of people   =   0.5 × 66   =   33

33 people said they would vote Liberal Democrat.

Example

The pie chart below shows the results from a survey of people in Scotland who were asked how they would vote in the next general election. If 33 people said they would vote Labour, and the angle for that slice of the pie is as shown, find the total number of people surveyed.

Scotland

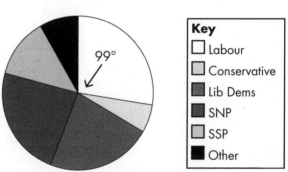

Key
- ☐ Labour
- ☐ Conservative
- ■ Lib Dems
- ■ SNP
- ☐ SSP
- ■ Other

33 people said they would vote Labour. The angle for the Labour slice is 99°. So we can write:

99° = 33 people

Now we can find the number of people represented by 1° of the pie chart.

1° = 33 ÷ 99   =   0.33 ... or $\frac{1}{3}$ people (again, don't worry about the absurdity of this answer!)

The total at the centre of the pie is 360°. This represents all the people surveyed, so we can work out the total number of people as follows:

Total number of people (360°)   =   0.33 ... (or $\frac{1}{3}$) × 360   =   120

A total of 120 people were surveyed.

Analysing data

Statisticians use a range of analytical methods to help them interpret data. Most are beyond the scope of this book, but we will look at two methods that we can use to gain a better understanding of our data.

Averages

Averages can be a useful way of interpreting data. An average is a number that gives a representative idea of a set of data.

Example

Mary and Peter have been asked to take a group of children on a day trip to the seaside. Mary is concerned because she doesn't know the ages of the children. All that she has been told is that the average age is 6.

The following table shows how many children of different ages are in the group.

Age	1	2	3	4	5	6	7	8	9	10
Number of children	0	1	1	4	5	4	3	5	1	1

Would Mary be right to assume that the average age of 6 represents the group?

Before we can answer this question, we need to know a bit more about averages. There are 3 types of average: the **mean**, the **mode** and the **median**.

 The **mean** is the average that people are usually familiar with. It is sometimes called the **arithmetic average** or **arithmetic mean**.

To find the mean of two or more values, add the values together and then divide the total by the number of values.

Example

Find the mean of the numbers 2, 5, 4, 9, 3, 2, 4, 10, 1, 4, 0.

Mean   =   (2 + 5 + 4 + 9 + 3 + 2 + 4 + 10 + 1 + 4 + 0) ÷ 11

(The sum of all the numbers is divided by 11 because there are 11 numbers.)

Mean = 44 ÷ 11 = 4

The mean of these numbers is 4.

 The **mode** of a set of values is the value that appears most often in the set.

Example

Find the mode of 2, 5, 4, 9, 3, 2, 4, 10, 1, 4, 0.

The number that appears most often is 4
So the mode of these numbers is 4.

 The **median** of a set of values is the middle value of the set after the values have been put in order, smallest to biggest.

Example

Find the median of 2, 5, 4, 9, 3, 2, 4, 10, 1, 4, 0.

First put the numbers in order (it can be helpful to cross out the numbers as you work through them so that you don't miss any out or use one twice):

0, 1, 2, 2, 3, 4, 4, 4, 5, 9, 10

The middle value is the value with equal numbers on either side:

0, 1, 2, 2, 3, 4, 4, 4, 5, 9, 10

↑

middle value

The median of these numbers is 4.

 If the set has an **even** number of values (a number that can be divided by 2) then there will be no clear middle value. If this is the case, take the *two* middle values, add them together, and divide the answer by two. This will give you a single middle value.

In all three examples we have used the same set of numbers, and the mean, mode and median have worked out to be the same. This is not always the case – often they will be different.

Let's return to Mary's and Pete's day trip to the seaside, and work out each type of average for the data.

Age	1	2	3	4	5	6	7	8	9	10
Number of children	0	1	1	4	5	4	3	5	1	1

To make it easier to work out the mean, mode, and median of the data, first write the data in a different form:

2 3 4 4 4 4 5 5 5 5 5 6 6 6 6 7 7 7 8 8 8
8 8 9 10

This represents 1 two year old, 1 three year old, 4 four years olds, 5 five year olds, and so on. Before continuing, compare this list of numbers with the table and make sure you understand the connection between them.

First we can work out the mean:

Mean = the sum of the ages divided by the total number of children

Total number of children = $0 + 1 + 1 + 4 + 5 + 4 + 3 + 5 + 1 + 1$ = 25

So, we will divide the sum of the ages by 25 because there are 25 children in total (found by adding up the numbers of children in the second row of the table).

Now we can calculate the mean by adding up the ages and dividing by 25:

Mean = $(2 + 3 + 4 + 4 + 4 + 4 + 5 + 5 + 5 + 5 + 5 + 6 + 6 + 6 + 6 +$
$7 + 7 + 7 + 8 + 8 + 8 + 8 + 8 + 9 + 10) \div 25$
= $150 \div 25$ = 6

The mean age of the children is 6.

Then we can work out the mode, by finding the number that appears most often. In our example, this would be the age belonging to the largest number of children. Looking at the table, we can see that there are five children aged 5 and five children aged 8. So in this example there are two modes: 5 and 8.

Lastly, we can work out the median. First we put the ages of the children in order from smallest to biggest. The middle number will then be the median.

2 3 4 4 4 4 5 5 5 5 5 5 6 6 6 6 7 7 7 8 8 8
8 8 9 10

↑

middle value

The median is 6.

So, which of these three averages would give Mary and Pete the most information about the ages of the group of children?

Probably the average that Mary was given was the mean, as this is the average that people use most in everyday life. However, it didn't really tell her much about the composition of the group. It might have made her think that most of the children were 6 years old, which wasn't true.

If Mary had been given the median average, it still wouldn't have given her much information, but she would have know that the 'middle' age of the group was 6.

If Mary had been told the mode, she would have known that there were more five-year-olds and eight-year-olds in the group than children of other ages. Knowing this might have been more useful to her.

 Instead of rewriting the data as we did above, we can determine the mean, mode and median directly from a table as shown below.

Age	1	2	3	4	5	6	7	8	9	10
Number of children	0	1	1	4	5	4	3	5	1	1

The mean is the sum of all the ages of all the children divided by the total number of children. To find the sum, instead of writing

(2 +3 +4 +4 +4 +4 +5 +5 +5 +5 +5 +6 +6 +6 +6 +7 +7 +7 +8 +8 +8 +8+8 +9 +10)

we can write

(2 × 1 + 3 × 1 + 4 × 4 + 5 × 5 + 6 × 4 + 7 × 3 + 8 × 5 + 9 × 1 + 10 × 1)

So we can calculate the mean:

mean = (2 × 1 + 3 × 1 + 4 × 4 + 5 × 5 + 6 × 4 + 7 × 3 +
8 × 5 + 9 × 1 + 10 × 1) ÷ 25
= 150 ÷ 25 = 6

Compare this with the slightly longer calculation shown earlier.

The mode is the age or ages that appear most often in the data. From the table we can see that there are 5 children aged 5 and 5 children aged 8. There are not as many children in any other age group. So the modes are 5 and 8

The median is the middle value.There are 25 numbers in the set, so the middle value must be the 13th value (12 values on either side). Looking at the table of values, we can count along the 'Number of children' row, adding as we go, until the total comes to 13 or more. Doing this, we can see that the 13th value must lie in amongst the sixes.The median value must be 6.

If finding the averages directly from a table in this way proves difficult, remember that you can always write out the data as a row of numbers as shown previously.

Averages aren't always very useful on their own. Next we will look at another statistic that might have been helpful to Mary and Pete, particularly if they had been given it as well as an average.

Range
The **range** of a set of values is the difference between the smallest and the largest value. It is a measure of the spread of the data.

Example
What is the range of the numbers 9, 35, 27, 4, 10, 28, 9, 38, 29, 5, 7, 8?

The range is given by:

Range   =   38 (the largest number) – 4 (the smallest number)   =   34

There is a difference of 34 between the smallest and the largest number.

Going back to the seaside trip, if Mary and Pete had been told the range of the children's ages then they would have known that they were going to look after a wide spread of ages.

Range of children's ages   =   10 (largest) – 2 (smallest)   =   8

There was an 8 year difference between the youngest and the oldest.

If Mary and Pete had been told this as well as the mean average, or the median average, they would have started to build up a clearer picture of what the group was going to be like.

Practice 1

1. What value is being indicated by the arrow in the following scales?

a)

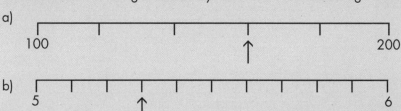

2. Look at the following bar chart of attendance at a doctor's surgery. How many people attended the surgery on Wednesday? How many people in total attended on Monday and Tuesday?

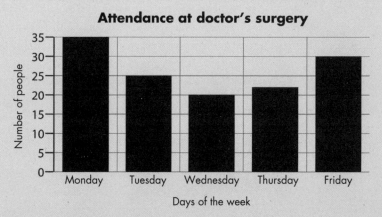

3. Draw a suitable bar graph to represent the following information about the colours of cars in a car park:

Car colour	Red	Black	White	Blue	Green	Yellow	Brown	Orange
Number of cars	24	18	5	20	8	4	6	5

4. Use the following line graph to convert 10°C to °F, and 95°F to °C.

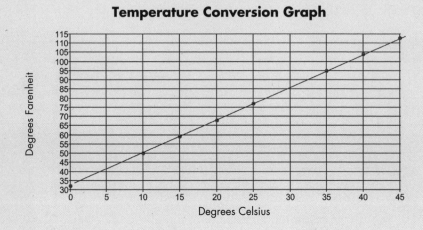

Temperature Conversion Graph

5. Draw a suitable conversion graph for changing litres to pints using the information in the following table.

Litres	4	8	12	16
Pints	7	14	21	28

6. The pie chart below illustrates a survey of 90 people who were asked to name their favourite holiday destination. Use the pie chart to calculate how many people said Greece was their favourite destination.

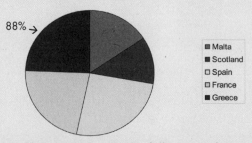

88% →

■ Malta
■ Scotland
□ Spain
□ France
■ Greece

7. Draw a suitable pie chart to represent the following information about someone's weekly budget:

Rent	Bills	Food	Travel	Clothes	Other
£100	£25	£90	£35	£20	£30

8. Find the mean, mode and median of the following numbers:

   7 7 5 0 4 7 6 7 2 2 1 2 2 2 5 7 5 3 5 2
   6 1 4

9. The following table shows the findings of a survey that asked people how many hours of television they watch each day. Find the mean, mode and median values for the hours of television watched.

Hours of television watched per day	0	1	2	3	4	5	6	7	8	9	10
Number of people	4	6	12	35	20	8	4	3	2	2	0

ANSWERS TO PRACTICE EXERCISES

Whole numbers

Practice 1 (page 8)
1. 44
2. 97
3. 174
4. 359
5. 856
6. 138
7. 977
8. 1288
9. 4899
10. 359989

Practice 2 (page 11)
1. 51
2. 91
3. 184
4. 460
5. 1003
6. 210
7. 1012
8. 2182
9. 7326
10. 124444

Practice 3 (page 15)
1. 15
2. 26
3. 54
4. 122
5. 445
6. 555
7. 4264
8. 2210
9. 3164
10. 22251

Practice 4 (page 17)
1. 19
2. 55
3. 4
4. 88
5. 245
6. 82
7. 188
8. 32
9. 367
10. 1178

Practice 5 (page 20)

1. 30
2. 42
3. 24
4. 45
5. 28
6. 72
7. 27
8. 49
9. 18
10. 25

Practice 6 (page 23)

1. 540
2. 368
3. 234
4. 1050
5. 3036
6. 7031
7. 1845
8. 18910
9. 282815
10. 20712

Practice 7 (page 25)

1. 4
2. 8
3. 3
4. 6
5. 7
6. 7
7. 6
8. 10
9. 7
10. 9

Practice 8 (page 29)

1. 247
2. 407
3. 26
4. 346
5. 62
6. 210
7. 3317
8. 231
9. 81
10. 1600

Practice 9 (page 33)

1. 11
2. 17
3. 96
4. 13
5. 123
6. 400
7. 69
8. 98
9. 36
10. 123

Answers to practice exercises

Practice 10 (page 35)

1. 80
2. 30
3. 80
4. 280
5. 1240
6. 300
7. 600
8. 1400
9. 12500
10. 1000

Practice 11 (page 39)

1. 14
2. 5
3. 5
4. 4
5. 10

Practice 12 (page 40)

1. 9
2. £19
3. £86
4. £4.80
5. £2000

Fractions

Practice 1 (page 46)

1. 7
2. 3
3. 9
4. 3
5. 24
6. 10
7. 15
8. 20
9. 15
10. 35

Practice 2 (page 49)

1. $\frac{4}{5}$
2. $\frac{2}{3}$
3. $\frac{3}{5}$
4. $\frac{8}{15}$
5. $\frac{3}{8}$
6. $\frac{3}{5}$
7. $\frac{4}{9}$
8. $\frac{4}{5}$
9. $\frac{3}{10}$
10. $\frac{41}{50}$

Practice 3 (page 52)

1. £45
2. £30
3. £540
4. £20
5. £63
6. £30
7. £90
8. £14

9. £9

10. £25

Practice 4 (page 53)

1. $\frac{2}{3}$

2. $\frac{1}{4}$

3. $\frac{9}{10}$

4. $\frac{15}{22}$

5. $\frac{9}{10}$

6. $\frac{1}{2}$

7. $\frac{3}{10}$

8. $\frac{2}{9}$

9. $\frac{3}{20}$

10. $\frac{1}{4}$

Practice 5 (page 55)

1. $\frac{11}{8}$

2. $\frac{29}{8}$

3. $\frac{11}{2}$

4. $\frac{20}{3}$

5. $\frac{11}{4}$

6. $\frac{47}{8}$

7. $\frac{21}{2}$

8. $\frac{27}{2}$

9. $\frac{38}{3}$

10. $\frac{201}{4}$

Practice 6 (page 56)

1. $2\frac{2}{3}$

2. $4\frac{1}{2}$

3. $3\frac{1}{3}$

4. $2\frac{1}{7}$

5. $3\frac{2}{5}$

6. $3\frac{2}{7}$

7. $5\frac{5}{9}$

8. $1\frac{1}{12}$

9. $17\frac{1}{7}$

10. $5\frac{1}{200}$

Practice 7 (page 57)

1. $\frac{3}{4}, \frac{2}{3}, \frac{1}{2}$

2. $\frac{7}{10}, \frac{3}{5}, \frac{1}{2}$

3. $\frac{3}{5}, \frac{12}{35}, \frac{2}{7}$

4. $\frac{3}{5}, \frac{2}{9}, \frac{1}{10}$

5. $\frac{7}{8}, \frac{2}{3}, \frac{5}{12}$

Practice 8 (page 61)

1. $\frac{2}{3}$

2. $\frac{3}{5}$

3. $\frac{1}{2}$

4. $\frac{5}{6}$

5. $\frac{19}{21}$

6. $1\frac{4}{15}$

7. $1\frac{7}{12}$

8. $1\frac{1}{6}$

9. $1\frac{11}{20}$

10. $1\frac{29}{60}$

Practice 9 (page 62)

1. $3\frac{3}{8}$

2. $8\frac{1}{4}$

3. $9\frac{5}{6}$

4. $13\frac{1}{24}$

5. $8\frac{1}{12}$

Practice 10 (page 63)

1. $\frac{5}{9}$
2. $\frac{1}{2}$
3. $\frac{7}{20}$
4. $\frac{5}{6}$
5. $\frac{11}{20}$
6. $\frac{1}{6}$
7. $\frac{16}{35}$
8. $\frac{19}{60}$
9. $\frac{31}{60}$
10. $\frac{1}{72}$

Practice 11 (page 65)

1. $1\frac{3}{4}$
2. $\frac{3}{4}$
3. $1\frac{1}{2}$
4. $5\frac{13}{24}$
5. $2\frac{7}{24}$

Practice 12 (page 66)

1. $2\frac{2}{5}$
2. $5\frac{5}{7}$
3. $1\frac{1}{2}$
4. 2
5. $3\frac{1}{9}$

Practice 13 (page 68)

1. $\frac{3}{8}$

2. $\frac{1}{2}$
3. $\frac{1}{6}$
4. $4\frac{1}{8}$
5. $25\frac{1}{3}$

Practice 14 (page 70)

1. 2
2. $1\frac{1}{2}$
3. 6
4. $\frac{6}{11}$
5. $1\frac{13}{14}$

Practice 15 (page 72)

1. £144
2. $\frac{3}{5}$
3. No
4. $\frac{1}{12}$
5. £12.50

Decimals

Practice 1 (page 80)

1. 47.06
2. 93.211
3. 10.794
4. 18.907
5. 606.2325

Practice 2 (page 81)

1. 2.12
2. 52.236

3.   4.16
4.   0.459
5.   9.207

Practice 3 (page 82)
1.   14
2.   43.61
3.   5.4
4.   575.5
5.   4.5

Practice 4 (page 85)
1.   3
2.   1.476
3.   3.6128
4.   3.075
5.   0.00000018

Practice 5 (page 87)
1.   2.08
2.   2.45
3.   0.12
4.   0.0008
5.   29.32875

Practice 6 (page 90)
1.   50.5
2.   1021
3.   0.12
4.   1531.3
5.   8

Practice 7 (page 93)
1.   0.32
2.   17.346
3.   4.83
4.   12.1
5.   0.209
6.   8.07
7.   18.004
8.   0.01
9.   251
10. 3.14286

Practice 8 (page 96)
1.   350
2.   1550
3.   80
4.   13
5.   0.62868
6.   0.306
7.   0.0006
8.   60.401
9.   0.01
10. 730

Money

Practice 1 (page 100)
1.   £108.35
2.   £476.98
3.   £40.06
4.   £91.55

5. £35.23
6. £691.52
7. £922.92
8. £1179.84
9. £196.44
10. £6.93

Practice 2 (page 101)
1. 450 euros
2. 1360 euros
3. 2160 dollars
4. 7875 kroner
5. 48 360 rupees

Practice 3 (page 102)
1. £245.90
2. £527.78
3. £2500
4. £157.14
5. £22.73

Practice 4 (page 105)
1. £8
2. £163.20
3. £133.98
4. £57
5. £5.25

Practice 5 (page 105)
1. £655.64
2. £380.36

3. £844.30
4. £1960.07
5. £913.50

Percentages

Practice 1 (page 110)
1. £5, £25, £50, £125, £250, £375
2. £6, £30, £60, £150, £300, £450
3. £10, £50, £100, £250, £500, £750
4. £1.24, £6.20, £12.40, £31, £62, £93
5. £0.52, £2.60, £5.20, £13, £26, £39
6. £2.50, £12.50, £25, £62.50, £125, £187.50

Practice 2 (page 112)
1. £200
2. £1500
3. £25
4. 180
5. £510

Practice 3 (page 114)
1. £28 080
2. 680
3. 156
4. £367.50
5. 4 910 151

Practice 4 (page 117)

1. 25%
2. 20%
3. 15%
4. 160%
5. 100%
6. 60%
7. 2.5%
8. 25%
9. 80%
10. 17%

Fractions, decimals and percentages – the connection

Practice 1 (page 122)

1. 0.375
2. 0.7
3. 0.8
4. 0.05
5. 2.25

Practice 2 (page 123)

1. $\frac{2}{5}$
2. $\frac{13}{50}$
3. $\frac{31}{250}$
4. $4\frac{3}{5}$
5. $12\frac{7}{50}$

Practice 3 (page 124)

1. $\frac{7}{20}$
2. $\frac{41}{50}$
3. $\frac{23}{50}$
4. $\frac{2}{25}$
5. $1\frac{1}{5}$

Practice 4 (page 125)

1. 35%
2. 40%
3. $37\frac{1}{2}$% or 37.5%
4. 30%
5. 175%

Practice 5 (page 127)

1. 0.62
2. 5.55
3. 0.02
4. 0.18
5. 1.42

Practice 6 (page 128)

1. 68%
2. 36%
3. 16%
4. 2%
5. 105%

Ratio and proportion

Practice 1 (page 132)
1. 1:4
2. 1:3
3. 1:4
4. 3:4:10
5. 1:4:8

Practice 2 (page 134)
1. 4:1
2. 20:1
3. 4:1
4. 3:1
5. 80:3

Practice 3 (page 135)
1. 2:5
2. 8:3
3. 3:8
4. 6:1
5. 15:4

Practice 4 (page 138)
1. £3000, £2000
2. £18000, £27000, £36000
3. 898g, 2694g
4. 240cm, 20cm
5. £280, £140, £70

Practice 5 (page 139)
1. £1000, £800
2. 12
3. 500, 300, 200
4. 12, 6
5. £56000

Practice 6 (page 143)
1. £18
2. 525g
3. £400
4. 90 minutes
5. 10.5 hours

Practice 7 (page 148)
1. 14km
2. 29cm
3. 3.2m
4. 40cm
5. 75m

Measurement

Practice 1 (page 156)
1. 34.5cm
2. 32.94m
3. 19.473km
4. 900000mm
5. 234.5m
6. 4000m
7. 237500cm

8. 2.837 km
9. 120.54856 km
10. 250 000 cm

Practice 2 (page 157)

1. 4.56 g
2. 20 kg
3. 456 000 mg
4. 500 g
5. 0.00045 kg
6. 2 000 000 mg

Practice 3 (page 157)

1. 34.5 cL
2. 19.87 L
3. 450 cL
4. 190 mL
5. 12.345 L
6. 320 mL

Practice 4 (page 157)

90 000	9000	90	0.09
82 000	8200	82	0.082
750 000	75 000	750	0.75
1 500 000	150 000	1500	1.5

Shape

Practice 1 (page 166)

1. 28 cm
2. 36 cm

3. 17 m
4. 17 m
5. 140 mm

Practice 2 (page 167)

1. 40 cm
2. 20 cm
3. 10 m
4. 36 m
5. 40 mm

Practice 3 (page 168)

1. 4 m 8 cm
2. 25 m 20 cm
3. 6 m 80 cm
4. 14 m 80 cm
5. 32 cm

Practice 4 (page 170)

1. 160 cm^2
2. 312 cm^2
3. 6.25 m^2
4. 10.5 m^2
5. 6400 mm^2

Practice 5 (page 171)

1. 2000 cm^2 or 0.2 m^2
2. 15000 cm^2 or 1.5 m^2
3. 6250 cm^2 or 0.625 m^2
4. 7000 cm^2 or 0.7 m^2
5. 6000 mm^2 or 60 cm^2

Practice 6 (page 172)

1. $14 \, m^2$
2. $70 \, m$
3. $3.6 \, m^2$
4. 3
5. 2

Practice 7 (page 178)

1. perimeter = $18 \, m$, area = $10 \, m^2$
2. perimeter = $90 \, m$, area = $284 \, m^2$
3. $536 \, cm^2$
4. $24 \, m^2$

Practice 8 (page 182)

1. $24 \, cm^2$
2. $120 \, cm^2$
3. $1.5 \, m^2$
4. $20 \, m^2$
5. $1200 \, cm^2$

Practice 9 (page 186)

1. $94.26 \, cm$
2. $7.855 \, m$
3. $942.6 \, mm$
4. $62.84 \, cm$
5. $565.56 \, cm$

Practice 10 (page 188)

1. $314.2 \, cm^2$
2. $7855 \, cm^2$
3. $227.0095 \, cm^2$
4. $7855 \, cm^2$
5. $28.278 \, m^2$

Time

Practice 1 (page 197)

1. 1520
2. 0705
3. 1835
4. 0759
5. 1000

Practice 2 (page 197)

1. 12.45 pm
2. 3.21 am
3. 5.23 pm
4. 2 pm
5. 12.46 am

Practice 3 (page 200)

1. 3 hours 24 minutes
2. 2 hours 4 minutes
3. 4 hours 30 minutes
4. 3 hours 38 minutes
5. 10 hours 50 minutes
6. 5 hours 15 minutes
7. 5 hours 26 minutes
8. 4 hours 49 minutes
9. 15 hours 27 minutes
10. 5 hours 27 minutes

Temperature

Practice 1 (page 204)

1. 15°C
2. 20°C
3. 25°C
4. 65°C
5. 0°C

Practice 2 (page 204)

1. 50°F
2. 95°F
3. 86°F
4. 59°F
5. 212°F

Data

Practice 1 (page 233)

1. a) 160   b)   5.3
2. 20, 60
3.

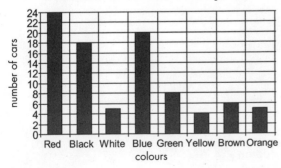

Colours of cars in a car park

4. 50°F, 35°C

5.
Conversion Graph

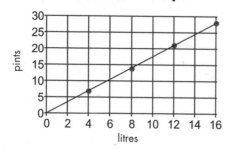

6. 22 people

7. Total budget = £300

Angles for each category:
Rent   120°
Bills   30°
Food   108°
Travel   42°
Clothes   24°
Other   36°

Weekly Budget

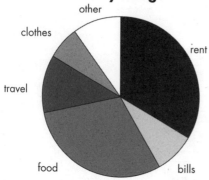

8. mean = 4, mode = 2, median = 4

9. mean = 3.5, mode = 3, median = 3

Appendix 1: Learning styles

About learning styles

Many people have a preferred **learning style** – a way of learning that suits them. Learning styles have been categorized and defined in a number of ways. One of the most widely used categorizations describes three main styles:

visual – 'watching'   auditory – 'listening'   kinaesthetic – 'doing'

For example, you might find that something is easier to understand if it is explained with the help of diagrams, pictures or video clips. That would suggest that you prefer a 'watching' or **visual learning style**. You might find listening to a tape of instructions helps you learn. This would suggest that you prefer a 'listening' or **auditory learning style**. You might find things easier to learn that allow you to have a 'hands-on' approach – things that you can 'have a go at'. This would suggest that you prefer a 'doing' or **kinaesthetic learning style**.

There are various questionnaires that will allow you to identify the style that suits you. You can find them on the Internet by searching for the words 'learning styles'. One of them is printed below, but before you investigate your preferred style by answering the questions, spend some time thinking about your learning and what you find helps you to learn. Just by doing that, you will probably be able to decide what kind of learner you are.

Two words of warning:

Learning styles questionnaires are designed to *indicate* a preferred style. They are not foolproof. If the results from the questionnaire indicate a style that you think is not your style, then trust your own judgement. You are the learner and you know what works for you. Sometimes the results will indicate that you have a preference for two out of the three styles. For some people the results will show that they are comfortable working in any style.

Also, the purpose of this type of learning style investigation is to make you more aware of how you learn, not to dictate how you ought to learn. In other words, even if you find that you seem to be a visual learner, that doesn't mean that you should refuse to work with 'listening' or 'doing' material! In fact, it is useful to try to develop your ability to work with other learning styles.

The following questionnaire and guide may help you to find learning strategies that suit you.

Appendix 1: Learning styles

Learning styles questionnaire

Choose the answer that most applies to you.

When you are reading, do you:

1. Make pictures in your mind?
2. Read the words out loud in your mind?
3. Imagine yourself doing the things described in the book?

When you are writing, do you:

1. Create a picture in your mind and find the words to describe this?
2. Make up sentences in your head and write them down as you say them?
3. Pace around the room, waiting for inspiration, or imagine yourself doing the actions in your writing?

When you are in a class or a group, which approach do you like best?

1. The teacher who uses a lot of visual aids.
2. The teacher who explains everything really carefully in words.
3. The learning session where you are encouraged to learn new skills by doing them.

When you are trying to learn information, do you:

1. Write down notes and try to remember them as a picture or link ideas with pictures?
2. Say your notes over to yourself, either in your head or out loud or on tape?
3. Pace around the room or have to practise skills repeatedly?

When you are trying to concentrate, do you:

1. Get distracted by pictures on the wall or in the book, untidy clutter on your desk, or in your room?
2. Get distracted by noises: for example, dripping taps, machinery hums, music?
3. Get distracted by movement around you?

How did you get on?

If you answered mostly **1**s, you learn best by **seeing** – you are probably a **visual learner**.

If you answered mostly **2**s, you learn best by **hearing** – you are probably an **auditory learner**.

If you answered mostly **3**s, you learn best by **doing** – you are probably a **kinaesthetic learner**.

If you answered a mixture of **1**s, **2**s and **3**s, you use a **mixture of learning styles**.

How to make your learning style work for you

Visual learners may see words in their heads. They tend to like orderly things, and they like to doodle. They like pictures, graphs or maps better than spoken directions. If you are a visual learner, you may find using mind maps helpful. You may also benefit from using coloured pens and paper, and working in an uncluttered environment.

Auditory learners like to sound out words when learning to spell, and rules when learning numeracy techniques. They learn best by listening, and may talk to themselves or move their lips while reading. They can be talkative and love discussion but need clear explanations. If you are an auditory learner, you will probably concentrate best in a quiet room. It may help you to talk about a topic in order to learn it. You may also benefit from using a tape recorder for learning material.

Kinaesthetic learners think best when moving, and often use gestures. They may find it hard to sit still. If you are a kinaesthetic learner, you may find it easier to learn numeracy rules or spelling if you trace words and symbols and practise writing them. You will find that you learn best by doing things and practising them regularly.

(Thanks to Jewel and Esk Valley College and Learning Connections for permission to use the questionnaire. Thanks also to Eleanor Symms, lecturer at Jewel and Esk Valley College, who adapted the material from original material published by the Learning Connections team at Communities Scotland.)

Appendix 2: Multiplication tables

2 times table

1 × 2	=	2
2 × 2	=	4
3 × 2	=	6
4 × 2	=	8
5 × 2	=	10
6 × 2	=	12
7 × 2	=	14
8 × 2	=	16
9 × 2	=	18
10 × 2	=	20

3 times table

1 × 3	=	3
2 × 3	=	6
3 × 3	=	9
4 × 3	=	12
5 × 3	=	15
6 × 3	=	18
7 × 3	=	21
8 × 3	=	24
9 × 3	=	27
10 × 3	=	30

4 times table

1 × 4	=	4
2 × 4	=	8
3 × 4	=	12
4 × 4	=	16
5 × 4	=	20
6 × 4	=	24
7 × 4	=	28
8 × 4	=	32
9 × 4	=	36
10 × 4	=	40

5 times table

1 × 5	=	5
2 × 5	=	10
3 × 5	=	15
4 × 5	=	20
5 × 5	=	25
6 × 5	=	30
7 × 5	=	35
8 × 5	=	40
9 × 5	=	45
10 × 5	=	50

6 times table

1 × 6	=	6
2 × 6	=	12
3 × 6	=	18
4 × 6	=	24
5 × 6	=	30
6 × 6	=	36
7 × 6	=	42
8 × 6	=	48
9 × 6	=	54
10 × 6	=	60

7 times table

1 × 7	=	7
2 × 7	=	14
3 × 7	=	21
4 × 7	=	28
5 × 7	=	35
6 × 7	=	42
7 × 7	=	49
8 × 7	=	56
9 × 7	=	63
10 × 7	=	70

8 times table

1 × 8	=	8
2 × 8	=	16
3 × 8	=	24
4 × 8	=	32
5 × 8	=	40
6 × 8	=	48
7 × 8	=	56
8 × 8	=	64
9 × 8	=	72
10 × 8	=	80

9 times table

1 × 9	=	9
2 × 9	=	18
3 × 9	=	27
4 × 9	=	36
5 × 9	=	45
6 × 9	=	54
7 × 9	=	63
8 × 9	=	72
9 × 9	=	81
10 × 9	=	90

10 times table

1 × 10	=	10
2 × 10	=	20
3 × 10	=	30
4 × 10	=	40
5 × 10	=	50
6 × 10	=	60
7 × 10	=	70
8 × 10	=	80
9 × 10	=	90
10 × 10	=	100

Appendix 3: Conversion tables

Temperature

Degrees Celsius (°C)	Degrees Fahrenheit (°F)	Degrees Celsius (°C)	Degrees Fahrenheit (°F)
0	32	21	69.8
1	33.8	22	71.6
2	35.6	23	73.4
3	37.4	24	75.2
4	39.2	25	77
5	41	26	78.8
6	42.8	27	80.6
7	44.6	28	82.4
8	46.4	29	84.2
9	48.2	30	86
10	50	31	87.8
11	51.8	32	89.6
12	53.6	33	91.4
13	55.4	34	93.2
14	57.2	35	95
15	59	36	96.8
16	60.8	37	98.6
17	62.6	38	100.4
18	64.4	39	102.2
19	66.2	40	104
20	68	100	212

Appendix 3: Conversion tables

Length

Imperial to metric

Imperial	Metric
1 inch (in)	2.54 cm
1 foot (ft) (= 12 in)	0.3048 m
1 yard (yd) (= 3 ft)	0.9144 m
1 mile (mi) (= 1760 yd)	1.609 km

Metric to imperial

Metric	Imperial
1 millimetre (mm)	0.03937 in
1 centimetre (cm) (= 10 mm)	0.3937 in
1 metre (m) (= 100 cm)	1.0936 yd
1 kilometre (km) (= 1000 m)	0.6214 mi

Area

Imperial to metric

Imperial	Metric
1 square inch (in^2 or sq in)	$6.4516 cm^2$
1 square foot (ft^2 or sq ft) (= $144 in^2$)	$0.0929 m^2$
1 square yard (yd^2 or sq yd) (= $9 ft^2$)	$0.8361 m^2$
1 square mile (mi^2 or sq mi) (= $3 097 600 yd^2$)	$2.59 km^2$

Metric to imperial

Metric	Imperial
1 square centimetre (cm^2 or sq cm) (= $100 mm^2$)	$0.155 in^2$
1 square metre (m^2 or square metre) (= $10 000 cm^2$)	$10.7639 ft^2$ $1.196 yd^2$
1 square kilometre (km^2 or sq km) (= $1 000 000 m^2$)	$0.3861 mi^2$

Capacity/volume

Imperial to metric

Imperial	Metric
1 cubic inch (in^3 or cu in)	16.387 cm^3
1 cubic foot (ft^3 or cu ft) (= 1728 cu in)	0.0283 m^3
1 cu yard (yd^3 or cu yd) (= 27 cu ft)	0.7646 m^3

Metric to imperial

Metric	Imperial
1 cubic centimetre (cm^3 or cu cm)	0.061 in^3
1 cubic metre (m^3 or cu m) (= 1 000 000 cm^3)	35.315 ft^3 1.308 yd^3

Imperial	Metric
1 fluid ounce (fl oz)	28.413 mL
1 pint (pt) (= 20 fl oz)	0.5683 L
1 gallon (gal) (= 8 pt)	4.5461 L

Metric	Imperial
1 millilitre (mL)	0.035 fl oz
1 litre (l) (= 1000 mL)	1.76 pt

Weight/mass

Imperial to metric

Imperial	Metric
1 ounce (oz)	28.35 g
1 pound (lb) (= 16 oz)	0.4536 kg
1 stone (st) (= 14 lb)	6.3503 kg
1 hundredweight (cwt) (= 8 st)	50.802 kg
1 ton (= 20 cwt)	1.016 t

Metric to imperial

Metric	Imperial
1 gram (g) (= 1000 mg)	0.0353 oz
1 kilogram (kg) (= 1000 g)	2.2046 lb
1 tonne (t) (= 1000 kg)	0.9842 tons

Appendix 4: Useful websites

There are many websites that allow you to practise your numeracy skills. You may wish to try the sites in the following list: some allow you to download exercises to practise at home, while others allow you to practise online. You may also find other useful sites by entering the word 'numeracy' into your favourite search engine.

http://www.bbc.co.uk/skillswise

http://www.ltscotland.org.uk/nq/coreskills/numeracy.asp

http://www.move-on.org.uk/practicetests.asp

http://www.bbc.co.uk/schools/gcsebitesize/maths/

http://www.fodoweb.ca/education.asp

http://bdaugherty.tripod.com/keyskills.html

Details of websites were correct at the time of publication but may be subject to change.

Index

Index